WAS THE WORLD TRADE CENTER DISASTER IN PROPHECY? SHOULD AMERICA CONVERT TO THE METRIC SYSTEM? DID THE BIBLICAL EXODUS REALLY TAKE PLACE? WHAT CAUSES LYME DISEASE? WHEN WILL THE ANTICHRIST APPEAR? WHICH IS CORRECT, EVOLUTION OR CREATION? IS THE LORD REALLY GOING TO RETURN TO EARTH? WHAT DO THE PROPHETS HAVE TO SAY ABOUT THE WORLD'S FUTURE?

TRAIL OF PROPHECY

First Edition

I0192568

By Edward Oliver

WAS THE WORLD TRADE CENTER DISASTER IN PROPHECY? SHOULD AMERICA CONVERT TO THE METRIC SYSTEM? DID THE BIBLICAL EXODUS REALLY TAKE PLACE? WHAT CAUSES LYME DISEASE? WHEN WILL THE ANTICHRIST APPEAR? WHICH IS CORRECT, EVOLUTION OR CREATION? IS THE LORD REALLY GOING TO RETURN TO EARTH? WHAT DO THE PROPHETS HAVE TO SAY ABOUT THE WORLD'S FUTURE?

A MESSAGE TO THE AMERICAN PEOPLE

We have been the recipients of the choicest bounties of heaven, we have been preserved in peace and prosperity, we have grown in numbers, wealth and power as no other nation has ever grown.

But we have forgotten the gracious hand which preserved us in peace, and multiplied and enriched and strengthened us, and we have vainly imagined, in the deceitfulness of our hearts, that all these things were produced by some superior wisdom and virtue of our own.

Intoxicated with unbroken success, we have become too self-sufficient to feel the necessity of redeeming and preserving grace, too proud to pray to the God that made us.

Abraham Lincoln

INTRODUCTION

WHERE IS ONE TO FIND security in such an insecure world? The love and kindness once available to us within the family fold, has now all but disappeared due to the demands of the modern working world. We can't even enjoy our rapidly disappearing outdoors unless we're willing to dress up in insect-proof clothing and saturate ourselves with copious quantities of poisonous insect sprays.

There also seems to be no comfort for us within the religious realm. All the world's greatest religions seem to agree that once in the very distant past, a man named Abraham walked the Earth and fathered a generation of people under the protection of an invisible God. Our spiritual salvation therefore rests in an invisible entity that we only get to meet after we've exhausted all our time here on Earth.

If this plight sounds familiar to you, then you may find this book somewhat intriguing. It's a collection of seemingly unrelated short stories, covering the many true events that acted together to bring us to the place we now occupy in the world today. Somehow we seem to have survived the threat of nuclear holocaust and successfully arrived at the seventh millennium of civilized man's existence here on Earth.

We all have questions about the world around us and how it operates. The answers to these questions are not always easy to come by. No one doubts that modern technology has

Introduction

brought us many comforts, but human problems have yet to be solved by the miracles of technology.

From the Adam and Eve event in 4000 BC, to modern space explorations of 2000 AD, we've had some 6000 years in which to create the perfect society; but the same old human problems still seem to plague us. I wonder; did we really learn anything in all that time?

Many of the events that helped to determine our destiny were first recorded for us in the Holy Scriptures. The Hebrew prophets who recorded this ancient knowledge also claimed to have received divine revelations concerning the future. The question of prophecy, and of our prophetic destiny, is something we will also explore in this book. For many years we've listened to those who interpret prophecy for us, and many times been frightened by their predictions. When these prophecies didn't come true, we often blamed the prophets themselves for the error. But is it the prophets we should blame, or just those who grew rich selling books and videos about their predictions? In this book we'll be taking another look at some famous prophecies to see if in fact prophecies may sometimes come true.

The history of the American people has been determined through the occurrence of many prophetic events. If we wish to know what lies in the future, we must first explore our past.

I hope you'll enjoy reading these strange, but true stories about our historic past, and I hope they will raise some questions in your mind concerning the role that prophecy may have played in our bumpy ride through history. Most of us live in the world of the seen, but constantly wonder about the world of the unseen. I also hope this book will raise some more questions in your mind about that unseen world.

Author

CONTENTS

CHAPTER ONE:
THE WORLD TRADE CENTER

ON THE MORNING OF SEPTEMBER 11TH, 2001, American Airlines Flight 11 left Logan Airport in Boston, bound for LA International Airport in California. It was a just another Tuesday for most of the passengers of Flight 11, just three more days until the weekend.

About one half hour into the flight, Boston air traffic controllers suddenly noticed that Flight 11 had turned around to a new heading. They desperately tried to contact the pilot of the aircraft to find out what was going on, but received no response.

The airplane, originally westbound, was now heading in a southerly direction toward New York City. Air traffic controllers immediately instituted standard emergency procedures, but no one knew what was going on aboard the plane.

When American Airlines passengers originally boarded Flight 11 at 7:45 AM that morning, they had no idea that they were in the company of a group of Arab terrorists who were planning to hijack the flight. Mohamed Atta and four other men were planning to take over the aircraft and crash it into the World Trade Center in New York City. Flight 11 had a total of 92 passengers on board, including the aircraft's crew and the hijackers.

At approximately 8:15 AM, the five hijackers took control of the airplane's cockpit and turned the aircraft around to a new course. The hijackers did not know how to navigate such a large jet, but they did know how to steer one, and intended to visually follow the Hudson River all the way to their New York City destination. They had trained for this mission for over a year, and had carefully planned every detail of the flight.

Later investigations would reveal that the hijackers had made a number of dry runs in advance of their deadly mission. James Woods, a Hollywood actor, had taken an earlier flight from Logan to LA International during one of these dry runs, approximately one month prior to the attack.

On that flight, Woods was sitting in first class accompanied by four Middle Eastern men who were behaving very strangely. They never drank or ate anything during the flight, and were constantly talking to each other in subdued tones, taking notes, and periodically glancing out the windows of the airplane. Woods had the distinct impression that these men were preparing for a future hijacking of this flight. He reported his suspicions to airport authorities when he landed in LA, and also notified the FBI.

These dry runs proved to be very valuable to the hijackers. It was later found out that on one flight, the hijackers gained easy access to the pilot's cabin by telling the pilot that they were flight students and wanted to see what a jet cockpit looked like. They were also able to calculate the exact timing of the hijacking to coincide with the plane's passing over New York's Hudson River Valley. The hijackers planned to follow the Hudson River all the way to New York City.

On the morning of September 11th, the citizens of New York were in a state of shock when an airplane suddenly crashed into the north tower of their World Trade Center. Workers in the south tower called home to reassure their families they were OK, and would soon be evacuated, so New York firefighters could deal with this terrible accident.

The World Trade Center

News bulletins flashed across television screens all over the world, as live coverage of the incident took over normally scheduled programming on all major networks. Viewers all over the globe witnessed a flaming hole in the side of the World Trade Center north tower where an aircraft had crashed into it.

When it was announced that the plane was a large commercial jet from Logan Airport, instead of a smaller private plane, some people began to wonder whether this was truly an accident. Then, before their thoughts could congeal, another plane suddenly smashed into the Trade Center's south tower, sending fire and debris crashing down into the streets below.

Now, suddenly, the unthinkable became a reality. The United States was under attack! Attention soon broadened to cover the rest of the nation. What else might follow? Everyone held their breath in expectation of what was now unfolding in front of their eyes.

Suddenly, another report from Washington DC; the Pentagon has been struck by an airplane, and the White House might also be in danger. A fourth plane has also been reported missing. Things calmed a bit as news reports seemed to indicate that these attacks were confined to the hijacking of commercial jetliners. All air traffic in the continental United States was immediately halted at all major airports in order to deal with the situation.

The World Trade Center towers were apparently designed to withstand the impact of aircraft, since small planes had crashed into tall buildings before, but the towers were not specifically designed to withstand the impact of large jet airliners. Such crashes were considered nearly impossible, since commercial jets were so closely monitored by radar.

The superstructure supporting the towers was unable to withstand the intense heat from the large quantities of jet fuel, and ultimately gave way when heat weakened the building supports. Luckily, many people inside the World Trade Center towers were able to escape the disaster. Even some workers on the uppermost floors successfully escaped to the streets below. Loss of life in the Trade Center catastrophe however, was still one of the largest in US history. The shape of the New York City skyline, and the security of the American people, were forever altered by this tragic event.

Shortly after the World Trade Center attack occurred, the Internet was overwhelmed by claims that this attack had been foretold by the prophecies of Nostradamus. Poems allegedly written by Nostradamus were circulated on the Internet, describing the horrible event in great detail. These poems had been hurriedly pieced together from many different Nostradamus quatrains by Nostradamus fans desperately trying to find a verse to fit the occasion. Most Americans didn't pay much attention to these wild claims, since Nostradamus prophecies were quoted almost every time a major world disaster occurred.

Writing over four centuries earlier, Nostradamus had in fact produced some very interesting predictions of future events. One notably accurate prediction concerned the famous French scientist Louis Pasteur. In Chapter 1, Quatrain 25, of Nostradamus' famous book the *Centuries*, he not only included Pasteur's name, but foretold Pasteur's discovery of the microbe, and described the ridicule Pasteur suffered at the hands of his peers in the European medical community. The truly incredible part of this famous prophecy was that it was recorded over two centuries before Pasteur was born.

Nostradamus was a Jewish physician and prophet who claimed to have received hundreds of visions about future events. He was born Michel de Nostradame in the St. Remy province of France in the year 1503. He was the eldest of five sons, and his father was a notary. His family had recently converted to Christianity in order to avoid persecution by the Holy Roman Church of Europe. Nostradamus received his Bachelor's degree from Mont Pellier medical university in France, and later earned a PHD.

In 1538, an epidemic of bubonic plague killed his entire family, and he wandered Europe for many years tending to the sick and destitute. In the year 1555 he published his famous book the *Centuries*, a collection of prophecies on future events. He recorded his visions in the form of four-line poems known as quatrains. His book contained 10 chapters of 100 quatrains each. Chapter 7 of the book was never finished, and contains only 42 quatrains. The quatrains are not recorded in any understandable order, so dating them is very difficult.

Nostradamus also very cleverly disguised his quatrains in order to confuse those attempting to interpret the prophecies before their fulfillment. Nostradamus' quatrains are hand-written in Old French, and thus very difficult to translate. Nostradamus was also known to anagram certain words or names in order to disguise them. He would sometimes also include a "punch" line in certain of his quatrains which, when correctly translated, revealed the true meaning of the poem. Through these clever deceptions, Nostradamus was able to keep his prophecies successfully veiled for many centuries.

Among Nostradamus' many poems, was one that specifically mentioned towers that were "shaking" and "on fire" in New York City. This poem, recorded in Chapter 1 of the *Centuries*, was Quatrain number 87, and specifically mentioned that the event would take place in New York City. This quatrain generated quite a bit of interest in the world of prophecy for many decades. When the basement of New York's World Trade Center was bombed in 1993, most Nostradamus fans thought this was the fulfillment of the prophecy, and therefore retired the quatrain from their list of active prophecies. But when the World Trade Center was attacked for a second time on September 11th, 2001, interest in the prophecy was suddenly renewed. The last line, or "punch" line, of the poem had never been successfully translated. A rough English translation of the prophecy yields the following result:

CHAPTER 1, QUATRAIN 87 (English translation)

Great symbols on fire in the center of the mainland

Will cause trembling in the towers of New York City.

Two great skyscrapers will be continuously attacked,

This is when Arethuse turn-around to a new course.

In this quatrain, Nostradamus refers to the World Trade Center towers as the "Enno-sigee" (*enno* – great one(s), *sigee* – standing silent,) or great ensigns of the New York skyline. The World Trade Center towers were indeed the great ensigns, or symbols, of American capitalism, and were repeatedly referred to as such during media coverage of the disaster.

Another key word in this prophecy was the French word "Arethuse." It's exact spelling is very difficult for us to determine, since the original prophecy was hand-written in Old French, but the word "Arethuse" does in fact exist in the modern French language. It is a class of naval vessels and submarines. Some prophecy interpreters therefore thought that someone might hijack such a vessel, or submarine, and use it to attack New York City.

Another possible spelling for the word "Arethuse" is "Aerthuse." This word in Old French simply means "air-tubes" ("air" from the Old World "aer" and "tube" from the Old World "thuse," a surveyor's metal sighting tube used for measuring land). Could it be that Nostradamus' use of the word "air-tubes" was actually an attempt to describe the tube-shaped metal vessels in his vision that had the ability to fly through the air? If the word "aerthuse" does in fact mean "airplanes," then the prophecy would suddenly make sense.

But hold on! This would mean that Nostradamus was actually able to see through time! That's impossible of course! All modern scientists and astrophysicists agree that it's impossible for men to see through time. The ability to see through time however, is indeed the basis for all prophecy, including Bible prophecy.

There has been a long-standing controversy going on for many centuries between the scientific and religious communities over this exact issue. The prophecies of Nostradamus have always been a thorn in the side of modern science.

Most scientists just can't understand how people living in a modern, technologically advanced society like ours, could still hold on to silly fantasies about men seeing into the future. In this world of technological miracles like space travel and cloning, how could anyone still believe in the prophecies of an obscure 16th century Hebrew prophet?

Nostradamus had often been accused of being vague in his prophecies, but Quatrain #87 seemed to be quite specific in its description of the World Trade Center event. Unfortunately the last line of the prophecy could not be accurately interpreted until after the event took place. As with all other Nostradamus prophecies, the prophecy was not recognized for what it was until after its fulfillment. Foreknowledge of the event therefore, could not be scientifically demonstrated.

Could it be possible that the fate of the World Trade Center was locked in destiny? That's silly, there's no such thing as destiny! We all determine our own destiny. In this modern scientific world, the idea that destiny is predetermined by some force beyond our control, should have fallen by the wayside long ago. No one believes in prophetic destiny anymore, in fact, very few people even believe in God anymore, and are actively trying to remove Him from the public arena. But how else are we to explain this uncanny prediction? Could Nostradamus' prophecies actually be true after all, or is this just another case of strange coincidence?

There are more than 900 prophecies in Nostradamus' famous collection. Most aren't dated, and so it's left to the individual to interpret the prophecies on his own. Most modern observers however, do not believe in the accuracy of Nostradamus' predictions, and no Nostradamus prophecy has ever been successfully interpreted before its occurrence.

We'll delve further into this question of Nostradamus' alleged ability to see into the future in some future chapters of this book. For now, there is one thing that cannot be denied, and that is the fact that the following quatrain is printed in numerous books, all of which were published many years in advance of the World Trade Center disaster. It would therefore be difficult for us to deny Nostradamus' foreknowledge of the event. You can still purchase many of these books in modern bookstores today. The original prophecy #1-87 was written in Old French, and so I've included a glossary of definitions for those of you unfamiliar with the Old French and Latin terms.

Chapter I, Quatrain 87 (Old French)

Enno-sigee feu du centre de terre.
Great-promontories on fire in the center of the mainland,

Fera trembler au tours de Cite' Nuefve.
Will make tremble the towers of the City of New York.

Deux grands rochiers long-temps feront la guerre,
Two great skyscrapers continuously will be attacked,

Puis are-thuse rou-gira nouveau fleuve.

The World Trade Center

This is when air-planes turn-around to a new course.

OLD FRENCH AND LATIN DEFINITIONS:

Are-thuse – (L. survey tube, vessel) air-vessels, airplanes
Centre – (F.) center, middle
Cite' Neufve – (O.F.) City of New York, (ninth empire)
Deux – (F.) two
Enno-sigee – (L.) great-promontories, ensigns, symbols
Fera/feront – (O.F. faire) will make, will be
Feu – (F.) burning, on fire
Fleuve – (O.F.) route, course
Grands – (F.) grand, great
La Guerre – (O.F.) war, warlike attack
Long-temps – (O.F.) long-time, continuously
Nouveau – (F.) new
Puis – (F.) then, (OF) this is when
Rochiers – (O.F.) rock monoliths, skyscrapers
Rou-gira – (OF) round-gyrate, turn-around
Terre – (O.F.) land, island, mainland
Tours – (O.F.) towers
Trembler – (F.) tremble, shake

Prophecy #87 in Chapter 1 of the *Centuries* is not the only Nostradamus prophecy specifically mentioning New York City. Nostradamus mentions many world cities in his prophecies, but New York City is the only US city Nostradamus mentions. He refers to New York as "Cite' Neufve," or the "New City" (of the Ninth empire).

New York has always been the most important city in the US for many reasons. It is the number one commercial, financial, and entertainment capital of America. Back when our nation was first formed, New York City was proposed to be our nation's capital, but New York was not a very elegant city at the time, and eventually lost out to the more stately Washington DC area.

New York City however, is still the most important city in America. It's the home of the New York Stock Exchange, the United Nations, the New York Times, the New York Stock Market, and countless other famous institutions. New York City is also still the main business and trading center of our nation.

Nostradamus followers have for many years been trying to figure out how his prophecies would affect the United States. Prophecies mentioning New York City therefore were always viewed with great concern by most Nostradamus followers.

Nostradamus' prophecies were first translated many years ago, and unfortunately these original translations were not very accurate. No significant changes in the translations have occurred over the years, and so most Nostradamus books published since that time have merely carried through these original texts unchanged. The overall accuracy of these English translations is therefore quite poor; so many modern Nostradamus followers must labor under the disadvantage of not having accurate interpretations of his prophecies to work with. The Old French meanings have also changed quite considerably over the last 450 years, and often no longer directly relate to modern French terms.

Nostradamus followers are certain however, that "Cite' Neufve" does indeed refer to New York City. The term "Cite' Neufve" is used only 3 or 4 times in Nostradamus' book the *Centuries.* Prophecy number 87, involving the World Trade Center, was originally perceived to be an attack upon New York City's skyscrapers, but there are more prophecies that have also been interpreted as describing attacks upon the city.

Erika Cheetham, in her famous book *The Man Who Saw Tomorrow – The Prophecies of Nostradamus,* published in 1974, mentions three specific prophecies that she says foretell attacks upon New York City. Prophecy #6-97, describing a bombing,

Prophecy #1-87, describing a possible attack upon New York's skyscrapers, and Prophecy #10-49, describing New York poisoned through its water supply. Ms. Cheetham also indicates that another prophecy, Quatrain #9-92, might possibly refer to New York as well. These predictions did not seem to be very plausible in 1974 when her book was first published, but now in the 21st century, with the advent of Arab terrorism, these kinds of despicable acts are no longer quite so implausible.

Could it be that God was trying to warn us about these horrible tragedies through the Hebrew prophet Nostradamus? If so, it appears that no one was listening. According to modern science, there is no way that Nostradamus could have predicted these events. Scientists firmly maintain that it's impossible for anyone to predict the future, and therefore consider the prophecies of Nostradamus to be utter nonsense.

Unfortunately the future is still unknown to us. No one has yet been able to unlock the secrets of the ancient prophets. All prophecy, including religious prophecy, has therefore been relegated to the status of myth and fantasy.

On September 11th, 2002, on the one-year anniversary of the 911 disaster, over 5000 people played the number 911 in the New York State Lottery. Believe it or not, 911 actually did come up that evening as the winning number.

Only time will tell whether the events foretold within the pages of the *Centuries* will ever come true, and liberate the Hebrew prophet Nostradamus from a rather ignominious fate.

CHAPTER TWO:
THE CYCLE OF THE
PRESIDENTS

He took them both, these two great men
That dark and peaceful night,
To be with Him above the clouds
In His eternal light.

Anonymous

DO YOU BELIEVE IN DESTINY? In our modern scientific world, the concept of prophetic destiny has now been relegated to the realm of mere fantasy. When amazing coincidences do suddenly present themselves, they are usually attributed to chance happenings, and no one becomes overly concerned about their significance. The scientific and religious communities have been arguing about the matter of prophetic destiny for many years, but the question has never been settled to anyone's satisfaction.

Some interesting examples of unusual coincidence are recorded for us in the history of the US presidency. The presidency of the United States has always been the subject of much attention by members of the media, who closely follow its many controversies and scandals.

The Cycle of the Presidents

The assassination of an American president is an event that always draws an enormous amount of attention from the media. The assassination of President John F. Kennedy in particular, was the subject of much media attention concerning conspiracy theories and the like. The thought that prophetic destiny might have had anything to do with the fate of any of our presidents has probably never crossed the minds of most people. In this short story however, we're going to take another look at that subject to see if in fact, prophetic destiny might have had anything to do with the Kennedy incident, or others like it.

President John F. Kennedy was probably the most popular president of modern times. His background was not the background of a typical American president. His father was Joseph P. Kennedy Sr., a Boston millionaire, and his mother was Rose Fitzgerald Kennedy, daughter of the late John "Honey Fitz" Fitzgerald, former mayor of Boston. Joseph P. Kennedy Sr. was a clever and ingenious businessman and politician, who was quite adept at advancing his power and authority within the Boston Irish community, and also within the Democratic Party.

The elder Kennedy had supported the presidential campaign of Franklin D. Roosevelt, and received the Chairmanship of the Securities and Exchange Commission in return for that favor. Roosevelt later appointed Kennedy to the office of Ambassador to Britain.

Joe Kennedy enjoyed wielding great political power, but did not particularly enjoy the attention and public exposure that went along with it. He much preferred to work behind the scenes, as the power behind the throne. Kennedy therefore conceived a plan to groom his eldest son, Joseph Patrick Kennedy Jr., for the office of President of the United States.

Kennedy held enough power and influence to successfully carry out this plan, but his plan was nearly thwarted when his son Joe Jr. was unexpectedly killed in an airplane explosion in 1944. Kennedy's favorite daughter Kathleen was also tragically

19

killed four years later in an airplane crash in Europe on the way to meet her father.

Kennedy then decided that his next eldest son, John Fitzgerald Kennedy, would be groomed for the office of President. After the war ended, Joe Kennedy used his considerable power and influence to run his son John for congressional office. The elder Kennedy took a terrible tongue lashing from the press however, when another man with the same name as John's opponent mysteriously appeared on the ballot. With dad's help, and his opponent's vote thus split, the younger Kennedy easily won the election.

Young Congressman John F. Kennedy went on to become a US Senator in 1952, and was finally elected to the office of President of the United States in 1960. Joe Kennedy's plans for John ultimately did come to fruition, but his own plan for seizing power was short lived. The elder Kennedy suffered a massive stroke on December 19th, 1961, and was left paralyzed and essentially speechless. He died November 18th, 1969, after living just long enough to see two more sons die tragically.

President John F. Kennedy was the first member of the Roman Catholic Church to hold the office of President of the United States. JFK was a liberal idealist who championed the twin causes of science and technology. The Kennedy years marked the beginning of a social and moral change in America that would continue on for many decades, and eventually alter the very fiber of American life. The previous reign of conservative White Anglo-Saxon Protestant presidents was now over. Kennedy however, had made many enemies during his brief tenure as President.

The assassination of President John F. Kennedy in 1963 resulted in one of the largest criminal investigations in US history. Kennedy was hated by many in the South because of his stand on civil rights. His efforts to eliminate segregation in the South stirred up feelings that had not been experienced since the Civil

War when President Lincoln had walked a similar path. Adlai Stevenson, sent on ahead as an advance man for Kennedy's trip to Dallas, was spat upon by angry southern white supremacists, and the stage was thus set for John F. Kennedy's date with destiny.

On November 22nd, 1963, President Kennedy was riding in an open limousine in a parade in Dallas, Texas, when assassin Lee Harvey Oswald shot him from the sixth floor window of a book warehouse located along the parade route. Oswald fired three shots at the president. The first shot struck the president in the neck, and the second shot, which proved to be the fatal shot, struck the president in the back of the head.

Oswald made the fatal mistake of running, instead of walking, away from the scene of the crime, and a description of him was immediately issued to Dallas police. Oswald shot Dallas police officer J. D. Tippit while fleeing, and was tracked to the Dallas Theater where he was promptly arrested for the crime.

Oswald was taken to Dallas Police Headquarters for questioning concerning the shootings, and Dallas police decided that they had enough evidence to hold him. Threats on Oswald's life were phoned in to the police department and police therefore decided it would be necessary to move Oswald to a more secure facility.

The next day, while police were attempting to transfer him out of the police station, Oswald was shot and killed by Jack Ruby, a local night club owner. Ruby gunned down Oswald in the basement of the Dallas police station.

Rumors about a conspiracy were running rampant, and Oswald, the most important source of information concerning any conspiracy, had now been silenced forever.

A special investigative commission was set up by the federal government to look into the circumstances surrounding the Kennedy assassination. Early into the investigation, federal investigators were struck by the many similarities between the

Kennedy assassination and the assassination of Abraham Lincoln that occurred a century earlier. Both presidents were shot in the back of the head, on a Friday, while seated next to their wives, and both assassins were southern white extremists.

Many other strange coincidences were also noted during the investigation, for instance,

President Lincoln was shot while sitting in Ford's Theater.
President Kennedy was shot while sitting in a Ford automobile, model Lincoln!

The Lincoln's maid, whose name was Mrs. Kennedy, had pleaded with him not to go to Ford's theater.
The Kennedy's maid, whose name was Mrs. Lincoln, had pleaded with him not to go to Dallas.

Lincoln's assassin shot him in a theater, and was captured in a warehouse.
Kennedy's assassin shot him from a warehouse, and was captured in a theater.

Needless to say, federal investigators were quite perplexed by all these strange coincidences, but these unusual coincidences proved to be just the tip of the iceberg. As investigators dug deeper into the case, more strange facts were uncovered. For instance, President Lincoln's vice president, Andrew Johnson, was the only vice president in history never to spend even one day in a schoolhouse, while President Kennedy's vice president, Lyndon Johnson, was a schoolteacher, and therefore spent nearly all his days in a schoolhouse.

All these things were just too strange to be mere coincidence, so investigators started digging deeper into the Lincoln assassination to see if there might be some sort of a weird conspiracy going on here, but the deeper they dug, the stranger things became. For instance, when Abraham Lincoln was first proposed as Republican candidate for the presidency, his proposed running mate was a former Secretary of the Navy whose name was, believe it or not, John Kennedy! And the man who protected Lincoln against the famous 1861 Baltimore assassination plot, was New York Police Superintendent, John Kennedy!

The parallels between these two presidencies were truly amazing, for instance:

Lincoln was elected President in 1860.
Kennedy was elected President in 1960.

Abraham Lincoln was elected to Congress in 1846.
John F. Kennedy was elected to Congress in 1946.

Vice President Andrew Johnson was born in 1808.
Vice President Lyndon Johnson was born in 1908.

Lincoln's assassin, John Wilkes Booth, was born in 1839.
Kennedy's assassin, Lee Harvey Oswald, was born in 1939.

Andrew Johnson died 10 years after President Lincoln.
Lyndon Johnson died 10 years after President Kennedy

Abraham Lincoln lost a son while in office.
John F. Kennedy lost a son while in office.

President Lincoln's assassin was shot before trial.
President Kennedy's assassin was shot before trial.

Lincoln's vice president was a former Democratic southern senator named Johnson.
Kennedy's vice president was a former Democratic southern senator named Johnson.

Trail of Prophecy

Abraham Lincoln is pictured on a US coin.
John F. Kennedy is pictured on a US coin.

The names Lincoln and Kennedy each contain 7 letters.

The names Andrew Johnson and Lyndon Johnson each contain 13 letters.

The names John Wilkes Booth and Lee Harvey Oswald each contain 15 letters.

Yes, it goes on, but because all these strange coincidences and many others were so popularized in books and magazines, many people are well aware of them.

What many people may not be aware of however, is the fact that these strange coincidences are only a small part of an even larger set of unusual coincidences surrounding the US presidency. Those who study prophecy are aware of some other very interesting facts surrounding the question of the destiny of US presidents. A closer examination of history reveals the following facts.

The president elected in 1840, William Harrison, died in office.

Twenty years later,

the president elected in 1860, Abraham Lincoln, died in office.

Twenty years later,

the president elected in 1880, John Garfield, died in office.

Twenty years later,

The Cycle of the Presidents

the president elected in 1900, William McKinley, died in office.

Twenty years later,

the president elected in 1920, Warren Harding, died in office.

Twenty years later,

the president elected in 1940, Franklin Roosevelt, died in office.

Are you beginning to notice a pattern here? American historians also noticed this 20-year pattern of US presidents dying in office. It was therefore with great interest that they viewed the 1960 election of President John F. Kennedy. The eyes of all historians were firmly fixed upon this young man to see if he would be the first president in history to end the mysterious 20-year death cycle of American presidents. A presidential assassination was considered to be a distinct possibility in this case, since President Kennedy was considered much to young to die of natural causes.

In 1956, author Jean Dixon predicted in a famous magazine article that the 1960 presidential election would be won by a Democrat who would be assassinated or die in office. When JFK was assassinated on November 22, 1963, he became the last American president to fall victim to this infamous 20-year presidential death cycle. This cycle that ran for 120 years, from 1840 to 1960, was now over.

Unknown to most observers however, this 20-year cycle of destiny, known to the ancient Greeks as the "Great Chronocrator," was overlain by an even larger 480-year cycle known as the Cycle of the Elements. The four elements involved in this larger 480-year cycle are Fire, Earth, Air, and Water, each element holding control of the cycle for 120 years.

The 20-year presidential death cycle that ran from 1840 to 1960 operated under the Earth portion of this Cycle of the Elements. The Earth Cycle, represented by the Earth element lead, meant that presidents elected under this 120-year portion of the cycle were destined to meet their fate through assassination, either by a lead bullet, or by lead poisoning.

In the 19th century, one popular method of killing someone was to mix lead arsenic sulfate into their mashed potatoes. After eating this mixture at a meal, the victim would suffer symptoms of severe food poisoning, and meet with an untimely death. A thorough investigation of presidential deaths occurring during this period has never been conducted, so we may never know if any American presidents died in this manner.

As previously mentioned, the 120-year Earth Cycle ran from 1840 to 1960. It was preceded however, by another 120-year cycle running from 1700 to 1820, known as the Fire Cycle. There were only two US presidential elections that occurred under the Fire Cycle (President George Washington did not obtain his office through public election). The two public elections occurring under the Fire Cycle were the elections of 1800 and 1820. Thomas Jefferson was the winner of the election of 1800, and James Monroe won the election of 1820.

President Thomas Jefferson's death, believe it or not, occurred on America's "Fire" holiday. Fire holiday you say? Yes, you know, that holiday we celebrate with things like fireworks, bonfires, sparklers, and other sorts of fiery objects. Thomas Jefferson actually died on the 4th of July, America's "Fire" holiday! Now there's a nifty coincidence for you!

It's actually a double coincidence however, because Jefferson was also the person who gave birth to the United States of America on this same date with his famous document, the Declaration of Independence, signed on July 4th!

But hold on, it's a triple coincidence, because Jefferson died on July 4th, 1826, which also happens to be the 50th birthday of the nation he gave birth to with his famous document. Read on folks, it gets even better!

President James Monroe, the second president to be elected under the Fire Cycle, was elected in 1820 and also died on the July 4th! That's right, a fourth coincidence! It was James Monroe's destiny to also die on America's "Fire" holiday since he too, was elected under the Fire Cycle.

Believe it or not, there's a fifth coincidence as well. While drafting America's Bill of Rights, President John Adams wrote a letter to Thomas Jefferson, who was in Paris at the time, asking him what specific rights he thought should be included in America's new bill. Jefferson sent back a list of individual freedoms that he thought should be included, and America's Bill of Rights was born.

Jefferson and Adams, through their two famous documents, the Declaration of Independence, and the Bill of Rights, are the two men most responsible for the precious God-given freedoms we all enjoy today. These God-given rights that all Americans hold against government oppression are what make our nation unique from all other nations on Earth.

Since Jefferson and Adams were the true founding fathers of the United States of America, God apparently decided to honor them in a very special way. Believe it or not, He took both Jefferson and Adams from us on the same day, July 4th, 1826, which also happens to be our nation's 50th birthday. Jefferson died at his home in Monticello, Virginia, and Adams died at his residence in Quincy, Massachusetts. Neither knew of the other's passing.

The Cycle of the Elements has now completed both its Fire and Earth cycles, and now operates under the Air Cycle. These strange cycles that seem to govern the destiny of America's presidents were originally identified more than 2500 years ago, at a time when the prophet Daniel ruled over the wise men and astrologers of the ancient city of Babylon (Dan. 2:48).

Trail of Prophecy

The Cycle of the Presidents is a haunting example of the many unusual events that helped to shape the destiny of the American people. Most people attribute these events to mere coincidence, but there are others who see a much deeper significance in their occurrence.

CHAPTER THREE:
THE EXODUS

And He said unto Abram, know of a surety that thy seed shall be a stranger in a land that is not theirs, and shall serve them; and they shall afflict them 400 years. And also that nation whom they shall serve, will I judge.

(Gen. 15:13)

WHEN WE THINK OF EPIC BIBLE STORIES, one that often comes to mind is the biblical Exodus, with its vision of the waters of the Red Sea parting, allowing the ancient Hebrews to escape, and then the waters rushing back in to drown the Pharaoh's army. Many of these epic stories told in the Old Testament are thought to be allegorical in nature, so as to prove a point, and others are sometimes viewed as a combination of myth and fact. The Exodus story is viewed by most modern scientists and historians as a myth, without any basis in either scientific or historic fact.

In our modern scientific world, miracles like the parting of the Red Sea, and the waters of the Nile turning to blood, are just too difficult for most people to accept as true fact. Recent archeological discoveries however, are beginning to shed a new light upon the Exodus story. It may now be time for us to take another look at the Exodus, to see whether or not it may have in fact occurred just as the Bible describes.

We must first state that biblical dating is not an exact science. Dates back to the time of Christ are fairly accurate, but dates prior to that time are prone to many inaccuracies. Most modern archeological evidence now places the date of the biblical Exodus at approximately 1470 BC. In order for us to accurately assess the Exodus event, it will be necessary to further explore the background of those times.

The 15th century BC was a barbaric time for mankind. Blood sports were a common form of entertainment, slavery was rampant, and morality practically non-existent. In 1470 BC, the barbaric Minoan civilization was thriving on the islands of the eastern Mediterranean Sea. The Minoans were one of the civilizations preceding the Greeks. They were a seafaring people who often traded with the Egyptians.

The Minoans controlled most of the sea trade in the Mediterranean area. Their ships sailed between Europe and Africa, and traded slaves and precious cargoes for money among the many civilizations of the region. Minoan society gradually spread its influence over the entire Mediterranean, and proved to be a dominant force in the region.

The Minoan culture was centered around money and pleasure. The Minoans worshipped the Golden Heifer. This form of cattle worship was very popular in those times, and spread rapidly throughout the Mediterranean. Exodus Chapter 32, Verse 4, and Deuteronomy Chapter 9, Verse 16, both tell us that worship of the Golden Heifer had even spread to the ancient Hebrews of Egypt. This form of cattle worship still survives today in India, giving rise to the familiar English expression, "holy cow."

The modern sport of bullfighting also owes its roots to the ancient Minoans. The Minoans practiced a form of bullfighting that differs from the bullfighting of today. In Minoan bullfighting, the bullfighter was an athlete who faced the bull without a cape. When the bull charged, the bullfighter would leap between the horns of the bull, and be vaulted over the bulls back in a graceful leap. This form of bullfighting was called bull vaulting. The bull vaulter had to be careful to time his leap just right, or he could be caught on the horns of the bull and gored to death, much to the delight of the crowds viewing the event. Such bloody and barbaric sports were very popular in Minoan society.

The Minoans populated many of the islands of the eastern Mediterranean. One of the more interesting islands they inhabited was the volcanic island of Stronghyli (Santorini). Stronghyli boasted many natural hot springs, and the island soon became a vacation paradise for wealthy Minoans. The island was the site of numerous palacial estates where the rich lived in great luxury, enjoying public baths and fresh crops of grapes and olives that could be grown year-round in the island's warm soil. Large, elaborately decorated barges, rowed by teams of African slaves, provided scenic tours around the island for its many wealthy visitors.

With so much material wealth, it was not long before Minoan society fell into a state of great moral decay. Paradise can often bring out the worst in human nature, and can sometimes also bring on the wrath of God.

While the Minoans were busy enjoying their island paradise, God's Hebrews were being held as slaves in nearby Egypt. The Hebrews had entered Egypt many years earlier at the invitation of Joseph, a Hebrew who'd been sold into slavery by his jealous brothers. Joseph was one of the twelve sons of the patriarch Jacob. When Jacob bestowed God's special blessing upon his son Joseph, Joseph's brothers could not control their jealousy, and sold Joseph into slavery in Egypt. It was Joseph's destiny however, to eventually rise out of that slavery and achieve a high position in Egypt.

Joseph was skilled in the interpretation of prophecy, and soon found favor with Egypt's Pharaoh by accurately interpreting a vision that Pharaoh had received. This vision foretold of a great famine that was about to occur in Egypt, lasting for 7 years. Joseph told Pharaoh that the Egyptians could survive the famine by storing away grain in the 7 good years prior to the famine's arrival.

Pharaoh was very pleased with Joseph's interpretation of his dream, and promptly placed Joseph in charge of preparing for the upcoming famine. Joseph was given full charge of Egypt's granaries, and proceeded to fill the granaries to overflowing in the 7 years before the drought.

When the drought and famine finally arrived, the Egyptians had plenty of food with which to feed their people. The ancient Hebrews living in the lands to the East however, were not so lucky. They were nomadic sheepherders who relied heavily upon the wild grasses of the field for their existence. When the famine arrived, they were totally unprepared, and in danger of starvation. When word arrived that the Pharaoh's granaries were filled to overflowing, the Hebrews decided to approach Pharaoh to beg for grain to sustain them through the drought.

The Hebrew patriarch, Jacob, sent his remaining sons into Egypt to beg for grain. When his sons arrived at Egypt's granaries, they did not know that the man they were begging from was their own brother, whom they had sold into slavery many years earlier. Joseph however, recognized his brothers, and ultimately revealed himself to them. He found it in his heart to forgive them for their treachery, and invited them into the land of Egypt as his guests. His father, Jacob, finally got to meet the son he thought he'd lost forever.

The Hebrews survived the famine, and prospered in Egypt. Over the many years of their stay in Egypt however, the throne eventually passed to a Pharaoh who did not know Joseph, and therefore felt no allegiance to Hebrews. The Hebrews soon found themselves working as slaves for their Egyptian masters. The iniquity of Joseph's brothers was thus repaid unto the fourth generation and beyond (Ex. 20:5). God's vengeance was delivered upon Joseph's brothers.

The Hebrew prophets foretold the birth of a Deliverer who would lead the Hebrews out of their Egyptian bondage. Pharaoh however, wanted no part of any Deliverer, and therefore ordered the death of all male Hebrew children born in Egypt. The wails of dying children and grieving mothers soon broke the stillness of many Egyptian nights.

A Levite woman named Jochabed gave birth to a male child, and set him adrift in the Nile River in a basket woven of reeds in order that his life might be spared. The Pharaoh's daughter found the child floating in the river and adopted him as her own son. She named him Moses (drawn-out), because she had drawn him out of the river. And so it was that God arranged for the Deliverer of the Hebrews to be raised up in Pharaoh's own house.

When Moses reached adulthood, he decided to explore his Hebrew roots. He made use of the blanket he was wrapped in as a child to find his family, for it was the custom of the ancient Hebrews to identify their clans in much the same way as the modern Scottish peoples of today. The Hebrews were also known to weave distinctive plaid patterns into their clothing in order to identify their various tribes. By using the plaid of the blanket, Moses was able to successfully locate his mother, Jochabed, and his brother Aaron, and sister Miriam.

Moses held a high position in Egypt, and often reviewed the work of the Hebrew slaves. One day, while reviewing a Hebrew work site, he caught an Egyptian taskmaster in the act of beating one of his Hebrew slaves. Moses killed the taskmaster and attempted to hide the body by burying it in the sand. Moses was exposed by the slaves for his act, and was forced to flee into nearby Midian in order to escape the wrath of Pharaoh.

Moses eventually married the daughter of the priest of Midian, and settled down to raise a family. But the Angel of the Lord appeared to Moses in a dream, and commanded him to return to Egypt to deliver his people from their bondage. Moses returned to Egypt and pleaded with Pharaoh to release the Hebrews from their slavery.

The pleading of Moses however, fell onto the deaf ears of Pharaoh, who refused to even consider releasing the Hebrew slaves. But the Hebrew God was quick to anger, and soon, from deep within the depths of the earth, a rumbling sound was heard beneath the volcanic island of Stronghyli. The great Stronghyli volcano, asleep for so many centuries, suddenly awakened and violently shook the Minoans' island paradise.

This great volcano, 7 times larger than any other in recorded history, was about to shake the great Mediterranean basin. Tens of thousands of Minoans desperately tried to flee their island in boats to escape the huge red clouds of volcanic ash generated by the great eruption. As the great volcano awakened, thick clouds of red smoke and ash rose up into the heavens and began moving in the direction of the Egyptian Nile delta.

These crimson clouds of ash drifted over the land of Egypt and settled into the headwaters of the great Nile River, causing its waters to turn a bright red. The fish in the river soon died from the acrid water, and millions of frogs were driven out of the river and onto dry land. The frogs poured into the homes of the Egyptians and perished by the thousands, creating a great stench throughout the Pharaoh's land.

Volcanic clouds of ash darkened the skies over Egypt for three full days, allowing millions of mosquitoes and flies to swarm during the daytime. Violent thunderstorms and hail soon enveloped the Nile Valley, smashing and ruining the crops of the fields. Animals that drank the poison river water soon died, and their carcasses drew swarms of flies. The flies also bit the Egyptian people, causing great boils to appear on their skin, spreading disease throughout the Pharaoh's land. Windstorms then brought huge clouds of locusts to devour the remainder of the crops in the fields and also the leaves of the fruit trees.

Exodus Chapters 7 through 10 tell us that in spite of all these plagues, Pharaoh still refused to release the Hebrew slaves. The Lord therefore instructed the Hebrews to slay a lamb and splash the blood of the lamb upon the lintels and door posts of their homes, for the Lord would slay the firstborn of any house without blood on its door posts. This was the root of the Passover, for the Angel of the Lord "passed over" the homes of the Hebrews.

That night the Angel of the Lord smote all the firstborn of the land of Egypt, including even the firstborn of the cattle. This was the final straw for Pharaoh, who finally yielded to the Hebrew God, and told Moses that he and his Hebrews should leave Egypt quickly.

The Hebrews, at the Lord's direction, had borrowed huge sums of money and jewelry from wealthy Egyptians just before they left Egypt. When this plot was revealed, Pharaoh flew into a great rage and sent the Egyptian army to chase down the Hebrews and kill them all. Moses knew that Pharaoh would be angry, and as he and his Hebrews embarked upon their Exodus across the desert, they decided to travel by both day and night to keep ahead of Pharaoh's troops.

It was difficult to travel in the Egyptian desert at night without the sun to determine direction, so Moses used the volcano as his guide. Exodus Chapter 13, Verse 21, and Nehemiah Chapter 9, Verse 19, both tell us that the Hebrews used the visible column of smoke from the volcano by day, and its plume of fire by night, to find their way across the Egyptian desert.

When the Hebrews finally arrived at the Nile River delta, they had to find a way to cross the canal that in those days connected the Red Sea and the Mediterranean Sea. There were only a few boats available, and it would take many days to cross the canal in this manner.

The Hebrews feared they would be trapped and killed by the Pharaoh's pursuing army, but just then, the sound of a gigantic explosion shook the Mediterranean basin. The volcanic island of Stronghyli, after erupting for many days, suddenly exploded and sank beneath the sea. The huge collapse sent a giant wall of water moving out in all directions from the spot that had once been the great Stronghyli.

A few minutes later on a nearby island, the inhabitants noticed the sea level dropping suddenly, as if the tide were going out. Older fishermen on the island had heard the great explosion, and knew from experience that the sea-water was being drawn out to fill the volume of an approaching tidal wave. They immediately shouted a warning for everyone to get to high ground as quickly as possible. Frightened mothers scooped up their children and ran as fast as they could for the center of the island. Time was short, for the tidal wave would arrive within a few minutes.

Ships on the open sea were unaffected by the wave because tidal waves travel over open ocean as gentle swells of water only a few feet high. They travel at hundreds of miles per hour, and begin to draw water out from the shoreline well in advance of their arrival. It is only when these waves reach the shallow waters near shore that they rise up to become the great terror we know as the tidal wave.

As the Stronghyli tidal wave approached each island, it rose up over 300 feet into the air like a giant cobra, and struck the shoreline, destroying fishing villages and smashing ships and small boats against the rocks. Striking island after island, the great wave destroyed everything in its path. The wave continued on across the Mediterranean in the direction of the Egyptian Nile delta, which unfortunately had no high shores to protect it.

The Exodus

Less than an hour after the great explosion of Stronghyli, the waters of the Red Sea canal were suddenly being drawn out into the Mediterranean Sea by a strange tidal flow. Egyptian villagers were frightened by this strange force that was suddenly emptying their canal, but Moses knew who had sent this, and ordered his people to be ready to cross the dry canal quickly. There was high ground on the other side where they could find refuge from what was about to happen.

Dust from the chariots of the approaching Pharaoh's army could now be seen in the distance. There was no time to spare. The Hebrews quickly gathered up their animals and belongings, and rushed to cross the dry canal bed in order to reach the safety of the opposite shore. From the top of the high plateau, they would at least have a chance of holding off Pharaoh's troops.

The Pharaoh's army charged into the dry canal bed in pursuit of the fleeing Hebrews, but when they had gotten only about half way across, they heard a loud roar coming from the north. When they turned to look, they beheld a sight that caused their hearts to stop. A great, thundering wall of water over 200 feet high was racing down the dry canal bed toward them. There was no time for retreat; before the troops could even turn their horses, the tidal wave was upon them. With heavy armor and weapons weighting them down, the Egyptians did not stand a chance against the giant wave.

The ancient Hebrews had a bird's eye view of this entire event from the safety of their position on the high plateau, east of the canal The Pharaoh's mighty army was totally destroyed in just a few moments right before their eyes. The rich lowlands of the Nile River delta were entirely submerged beneath the waters of the great wave. Hundreds of farming and fishing villages along the Mediterranean coast were totally destroyed in this great catastrophe.

Along with the story of Noah and the Flood, this story too was destined to become a permanent part of Hebrew literature.

In one momentous event, the Lord had destroyed an entire civilization, liberated another, and the great Pharaoh's army was no more.

The devastation caused by this cataclysm was greater than that of any other in recorded history. The Minoans were so traumatized by the event that they decided to relocate what was left of their civilization to the European mainland. There, the higher shores would prevent any possibility of a tragedy like this ever occurring again. It was many years before the Egyptians recovered from the damage to their farmland caused by the salt from the sea-water. For decades after this disaster, there was famine in the land of Egypt.

The story of the fate of the barbaric Minoan civilization eventually became a part of ancient Egyptian folklore. The story was passed down through the tales of many civilizations inhabiting the Mediterranean region. Many centuries later, the Greeks picked up the story from the priests of Sais, who recorded these great "cleansings" of mankind by the gods.

The Greeks promptly inflated the story by a factor of ten to make it worthy of Greek mythology, and the story eventually found its way down to the Greek philosopher Plato, who recorded it for us in his ten-book piece, The Republic. The story is still told today as the mythical tale of Atlantis.

To this very day, a bright layer of red volcanic ash still coats the Mediterranean Sea floor, testifying to the truth of this ancient tale. The path of this red ash leads directly from the remains of the volcanic island of Stronghyli, to the Egyptian Nile delta. The crimson cliffs of the island of Thera, and the other small islands surrounding the crater of the old Stronghyli volcano, still color the sea-water a bright red even today.

Archeologists are slowly beginning to uncover more scientific evidence of this ancient Bible event, that was until recently regarded as fiction. Science has now become a useful tool that modern man can use to reveal much about his historic past.

The Exodus

Many scientists no longer view the Bible as they once did. It has now become a useful source of information for scientists and archeologists alike, concerning the history of mankind, and man's prophetic destiny as revealed by God and His prophets. Now that the Seventh Millennium of man's civilized existence has finally arrived, scientists and archeologists everywhere are beginning to uncovering more information about the many prophetic events that seem to have governed the destiny of God's people.

CHAPTER FOUR:
THE GREAT PYRAMID

In that day, shall there be an altar to the Lord in the midst of the land of Egypt, and a pillar at the border thereof to the Lord.

(Isaiah 19:19)

IT IS STILL THE MOST AMAZING STRUCTURE on the entire face of the Earth, and although archeologists have been studying it for centuries, no one has yet been able to determine who built it, or what it stands for. It is the Great Pyramid, and the only thing we know for sure about it, is that it was not designed or built by the Egyptians. That's right, contrary to everything you learned in school, the Great Pyramid was not built by Egyptians.

Who built it then? In order for us to answer that question, it will first be necessary to review what modern science has thus far been able to reveal about this truly incredible structure. First of all, we know that the Great Pyramid has been standing in its present location for more than four thousand years. In spite of the fact that the pyramid is over four thousand years old, it still holds the world's record for being the most massive stone structure ever constructed by man. This fact alone makes it one of the most unique objects on Earth. Of the original Seven Wonders of the World, the Great Pyramid is the only one still in existence today.

The Great Pyramid

The Great Pyramid was originally constructed from over 2,000,000 hand-cut stone blocks weighing anywhere from 2 to 20 tons apiece. Every single one of these massive stone blocks had to be hand-quarried, cut into a perfect geometric shape, and then finely honed with such a degree of accuracy, that to this day a credit card cannot be fit between any two of them.

When the Great Pyramid was originally built over four thousand years ago, its exterior was completed covered in white polished limestone that shined so brightly, sailors often used the pyramid to chart their courses across the Mediterranean Sea.

In the second millennium before Christ, the Great Pyramid was truly a breathtaking sight to behold. It's polished limestone exterior made it appear as if an object from heaven had somehow mistakenly fallen to Earth. Around 1300 AD however, this beautiful limestone exterior was stripped off by Arab raiders who used the blocks to build their religious mosques.

Archeologists have since determined that the Great Pyramid contains no hieroglyphics. This is an unusual fact, for it places the structure in sharp contrast to all other pyramids in the area. Nowhere on the Great Pyramid are there any markings that would identify it as being Egyptian in origin.

In the late 1700's, European archeologists also discovered some other very interesting facts about this massive stone structure. They were able to determine through precise measurements, that the geometric shape of the Great Pyramid indicated its builders possessed an advanced knowledge of both mathematics and geometry. Archeologists uncovered the surprising fact that the Great Pyramid's height was in relation to the distance around its base, in the same proportion that the radius of a circle held to the circumference of a circle. This was a truly shocking discovery, for it clearly demonstrated that the pyramid's builders possessed an understanding of the pi relationship. It was not previously thought that any early civilizations were aware of the value of pi.

When this amazing discovery was first announced to the world, most European scholars absolutely refused to believe it. Teams of archeologists were immediately sent to the Middle East to further examine the Great Pyramid and take more precise measurements of its dimensions in order to dispel this wild theory.

When these archeologists attempted to determine the smallest common unit of measure used in building the pyramid, they found this measurement recorded on the top of a decorative boss located in the interior of the structure. This basic measuring unit turned out to be exactly equal to the modern day British inch! This shocking discovery presented archeologists with an even more difficult puzzle to explain. How could the British inch have possibly survived unchanged for more than four thousand years?

When archeologists sought to determine the next largest unit of measure used in building the pyramid, they found this measurement recorded in the pyramid's "Queen's Chamber," located at the 25th level of the structure. This measuring unit was found to be exactly equal to 25 British inches. Archeologists labeled this new measure the "Sacred Cubit" because the ancient Hebrew cubit was known to be about this length (actually 25.025025 inches). When archeologists measured the length of one side of the pyramid's base using this measurement, they were shocked to find out that it measured exactly 365.242 of these sacred cubits in length, representing the exact length of Earth's solar year! How could this be possible? This discovery sent yet another shock wave through the European scientific community. More expeditions were dispatched to Egypt to further investigate these claims.

European religious leaders were now beginning to express an interest in this ancient monument, originally built in biblical times. For the Bible had said that in the last days there would be an alter to the Lord still standing in the midst of Egypt (Isaiah

19:19). As with many other investigations into prophetic matters, it seemed that the deeper investigators dug into the matter, the more incredible their findings became. As more and more of these amazing revelations came to light, the fame of the pyramid steadily grew. Even the great Sir Isaac Newton expressed an interest in the pyramid. Newton had always held that the length of the ancient Hebrew cubit was somewhere between 23 and 26 inches. Newton was therefore quite anxious to find out whether the pyramid accurately recorded this ancient measure.

When it was finally determined that the pyramid set the ancient cubit's length at exactly 25 English inches, it was soon afterwards determined that this length also held a direct relationship to the size of the planet itself. The Sacred Cubit turned out to be exactly equal to one 10,000,000th part of the distance from the North Pole to the center of the Earth, meaning that at the time of the inception of this measurement, the Earth was exactly 500,000,000 inches tall! This shocking revelation meant that the builders of the pyramid were also aware of the exact size of the Earth itself. Their measurement was so precise that it even reflected the subtle effects of Post-Glacial Rebound, which slightly increased this measurement over time, due to the melting of the Earth's polar ice caps. How could all this be possible?

More archeological expeditions were immediately dispatched to Egypt to see if the Great Pyramid held any more revelations for mankind. More accurate measurements needed to be taken. As the development of man's technology advanced, more and more of the pyramid's secrets could be revealed.

Mapmakers had noticed that the Egyptian Nile delta extended into the Mediterranean Sea in the shape of a semi-circle. When the full arc of this circle was inscribed upon a map of Egypt, the Great Pyramid stood at its exact center! How could the ancient Egyptians have possibly determined this precise location without the use of modern mapping techniques?

Modern science had always taught that human evolution resulted in the advancement of mankind out of an ignorant and aboriginal past, but now the Great Pyramid was revealing past knowledge far in advance of that of the present. Could there have been a past civilization on Earth more advanced than our own? And what other secrets might this incredible repository for information contain for mankind?

It was not long before the world's greatest minds were focusing their full attention upon this massive stone object, that did indeed seem to be a source of previously unavailable information concerning the planet and the universe. Men like Charles Piazzi Smyth, Royal Astronomer of Scotland, and a host of others, undertook detailed studies of the Great Pyramid to see if any more information could be gleaned from it. They soon became convinced that this structure had been built by a source of inspired knowledge in the distant past, and that the pyramid had been specifically designed to serve as some sort of a time capsule for information on the history and the destiny of man. The amount of information available from the pyramid seemed to be limited only by man's ability to measure and interpret it.

As modern technology improved, the improved accuracy of measuring instruments allowed scientists to determine that the sides of the pyramid recorded not only the exact length of Earth's solar year, but also the anomalistic and sidereal years as well. It was also found that the two diagonal measures of the base of the pyramid produced a combined pyramid inch total of 25,827 inches, or the exact period of the Precession of the Equinoxes, recording Earth's 25,827-year wobble in space. This information was recorded again at the King's Chamber level of the pyramid. The distance around the exterior of the pyramid at the King's Chamber level is exactly 25,827 inches!

The pyramid's height when complete with its original tall capstone, was found to be equal to one $1,000,000,000^{th}$ part of the mean distance from the Earth to the sun (4500 years ago).

And the empty coffin inside the King's Chamber had an internal volume equal to that of the ancient English chaldron. The chaldron's quarter part is still recorded for us today in an English "quarter" of wheat.

Scientists also determined that the Great Pyramid was not only located at the exact center of the arc of the Egyptian Nile delta, but was also exactly oriented to the four cardinal points of the compass. It is also located at the exact point where all the land above that latitude exactly equals all the land below that latitude on Earth's northern hemisphere. The pyramid was then found to be located on the world's longest longitudinal and latitudinal land contact meridians as well. In other words, if all the world's present land masses originally broke away from one large landmass, as scientists now suspect, then the Great Pyramid is located at its exact center! The Great Pyramid also divided Upper Egypt from Lower Egypt, thus fulfilling the prophecy of Isaiah 19:19 that in the last days there would be an alter to the Lord standing in the midst of Egypt and also at the border thereof. All these revelations were more than modern science could rationally explain.

The mysteries of the Great Pyramid were truly mind-boggling. The floor of one of the interior anterooms for instance, is 116.26 inches in length, with 103.33 inches of it being constructed of red sandstone. This again records the magic relationship between the circle and the square, as the area of a circle with a diameter of 116.26 inches, is the same as that of a square whose sides are each 103.33 inches long.

The Great Pyramid sat for centuries without revealing any of its secrets to mankind, and no one was able to locate an entrance into the structure. Then, around the year 813 AD, the Arab Caliph, Al Malmoun decided to burrow into the north side of the pyramid to see if he could locate any inner chambers that might contain hidden treasure. Al Malmoun and his men had a very difficult time trying to chisel into the hard rock of the pyramid, but soon discovered that they could use fire to heat the rock, and then throw water on it, causing it to shatter. By using this method, Malmoun and his men were successfully able to tunnel about 150 feet into the north side of the pyramid, where they suddenly came upon a hidden interior shaft. This shaft then led them to a hidden staircase that led up into two secret chambers located deep within the structure. The uppermost of these two chambers contained a stone sarcophagus that today has been found to contain no trace of human remains. This upper chamber was dubbed the "King's Chamber" by archeologists.

The original 330-foot long passageway discovered by Al Malmoun and his men, led down at an angle from the exterior of the pyramid. It ended in a rough-hewn chamber located about 100 feet below the pyramid's base. A second passageway broke off from this passage at a point about 100 feet down, and led up into the pyramid's two internal chambers. This passageway then widened abruptly at a certain point into a much larger hallway known as the Grand Gallery before finally arriving at the entrance to the King's Chamber. Due to the design of the entrance to the King's Chamber, one had to bow down before entering that chamber.

Later investigations would lead to the discovery of a relationship between the length of these passageways and the chronology of the history of civilized man. It was discovered that, by using the equation of one inch per year, it was possible to trace the history of man down through the milleniums. Beginning at a point where the original shaft led in from the exterior of the pyramid, and assigning that point a value of 4000 BC, the second shaft broke off at the 1470 BC point, or the date of the biblical Exodus. This second shaft then expanded abruptly at the date of Jesus' birth, and accurately recorded his 33-year life here on Earth.

This passageway ended at the King's Chamber, representing roughly the 2000 AD point in the chronology of man's history.

The Great Pyramid

Archeologists surmised that these passageways traced the history of human civilization from its roots in ancient Sumeria in 4000 BC, to its end around the year 2000 AD. Many key events in the history of man did indeed seem to be marked on the inner walls of these passageways. Did all this mean that man would arrive at his final destiny around the year 2000? This theory was just too much for the scientific community to accept, and many archeologists supporting the theory were sharply condemned for their views.

The Great Pyramid still remains a source of controversy in the scientific world today. It stands in silent testimony to the existence of a past civilization far in advance of our own. The ancient Jewish historian Josephus recorded that the Great Pyramid was built by the children of Seth for the purpose of preserving knowledge. Josephus also related that the knowledge of the patriarchs was recorded in the tower of Babel as well, but God had destroyed the tower of Babel long before Josephus' time.

Could it be possible that the Great Pyramid really is proof of the existence of the biblical patriarchs, who lived for hundreds of years and attained great knowledge? There is just no explaining the strange enigma of the Great Pyramid, and scientists are still struggling to attempt to explain its existence. The Great Pyramid stands today as the most amazing structure on Earth, and no one has yet been able to satisfactorily explain its many mysteries.

CHAPTER FIVE:
WEST NILE VIRUS

And when you are gathered together within your cities, I will send the pestilence among you....

(Lev. 26:25)

IF YOU'RE THE KIND OF PERSON who is not satisfied with the fact that something happened, but also want to know why it happened, then you may find this story very interesting. It seems that in this modern world of medical miracles like organ transplants and cloning, we sometimes tend to lose touch with the basic realities that govern our existence here on Earth.

West Nile Virus has been the subject of many recent newspaper and magazine articles, and all sorts of opinions have been rendered on the causes and origins of this deadly disease that is slowly spreading across our nation. I therefore thought it appropriate for us to take a closer look at the history of West Nile Virus to see if we might reveal any clues as to its true origins.

In 1999, a few isolated cases of Eastern Equine Encephalitis were reported in certain communities along the New England coast. This disease was known to affect horses, but could also infect human beings as well. Since mosquitoes were known carriers of this deadly virus, public health officials immediately began a program of mosquito trapping in these areas to examine the mosquitoes for this deadly disease. Tests soon confirmed that mosquitoes were indeed carrying the virus.

Encephalitis is a disease that causes inflammation of the brain and spinal cord. It can affect almost any warm-blooded creature, but is especially dangerous to humans. In persons with suppressed immune systems, the disease can cause serious symptoms that include high fever, stiff neck, mental confusion, muscle weakness, coma, and in extreme cases, death. It is especially dangerous to people over the age of 50. Encephalitis is not a new disease and is found all over the world, but cases most commonly occur on the continent of Africa. In the fall of 1999, public health officials in New York City reported the first cases of West Nile Virus in the United States.

This deadly form of encephalitis was first identified in the West Nile region of Uganda many years ago, and can be transmitted to human beings by the bite of a mosquito. The first reported cases of West Nile Virus occurred in Africa in the year 1937, although it is likely that the disease existed for many years before that time. Since that original outbreak in Uganda, cases have been documented in many other nations as well. In the 1950's an outbreak was recorded in Egypt, and another outbreak occurred in some elderly patients in a nursing home in Israel in 1957. As the disease gained more recognition, new cases were reported in France and Egypt in the 1960's, and in South Africa in 1974. Romania recorded an outbreak in 1996, and Russia recorded a few cases in 1999. The disease is known to be hosted by migratory aquatic birds, including geese, and can be easily transferred to animals or humans by the bite of a mosquito.

Doctors initially had a difficult time trying to identify the virus. It was more than once confirmed through laboratory analysis to be Eastern Equine Encephalitis. It is quite common for such diseases to be misdiagnosed, even by the nation's top laboratories. It is only when a disease rises to epidemic levels that it is it given the attention necessary to arrive at a more accurate diagnosis. When this viral outbreak in the United States was finally classified as having epidemic potential, more accurate genetic tests were performed on it. Scientists were then able to determine that this strain more closely resembled the West Nile variety of encephalitis, and so it was agreed that this outbreak would be classified as West Nile Virus.

Since birds were known to host the disease, people were asked to bring in any dead birds found in affected areas. Crows and Blue Jays were brought in most often, since these birds are at the bottom of the food chain. Laboratory tests soon confirmed that many other species of birds were also carrying West Nile.

It is likely that there was West Nile Virus activity in the United States before these 1999 incidents, but the disease is very difficult to diagnose because its symptoms are so similar to those of the flu, and may include fever, headache, muscle aches, stiff neck and other flu-like maladies. It's possible that many people suffering from West Nile in the US were never properly diagnosed.

The first confirmed cases of West Nile Virus in the United States were recorded in New York City in 1999, where 62 people came down with the disease and exhibited its classic symptoms. These 62 cases resulted in 7 deaths. The death rate from diagnosed cases of West Nile Virus currently stands at approximately 6 deaths for every 100 serious cases. The number of West Nile Virus cases has risen rapidly each year as the disease spreads across the United States. The current North American case count as of 2003 is approximately 8000 serious cases, and will probably go much higher in the summer of 2004, as mosquitoes come out of their winter hiding places and begin to breed. Approximately 400 deaths have occurred in the United States so far as a result of these 8000 cases, and many people have also suffered permanent neurological damage from the disease. The disease seems to be most dangerous to the elderly, the very young, and persons with suppressed immune systems. The disease therefore has the potential to wreak havoc on the residents of nursing homes and elderly communities.

The first few cases of West Nile in the US were reported in New York City, but the following year many more cases were reported over a much wider area that extended from Connecticut to Maryland. By 2001, the disease had spread to most of the eastern seaboard, and new cases were being reported all the way from Maine to Florida. By 2002, West Nile Virus had successfully jumped the Mississippi River and more cases were reported from Michigan to Texas. The disease is still spreading westward, and by 2004, West Nile Virus promises to become a serious problem everywhere in the nation.

There is, as of this date, no effective human vaccine for West Nile Virus. There aren't even any effective treatments for the disease except for normal supportive therapies. It is therefore advisable for most people to try to avoid being bitten by mosquitoes.

The common North American House Mosquito is a known carrier of the virus, but it has recently been discovered that two new species of mosquito have invaded our nation and also become carriers of West Nile. These new species are the Asian Tiger Mosquito, and the Asian Japonicus Mosquito. These two new Asian invaders are also much more efficient at transferring the virus from one host to another, as much as 5 times more efficient! Both species have dark bodies with distinct white stripes, or scales, running along the sides of their abdomens. Be especially careful of these two new Asian invaders.

It is now feared that migratory birds are spreading West Nile Virus across our nation. Health officials have been experiencing problems recently with Canada Geese, due to a severe overpopulation of this species. Canada Geese populations have tripled in the last twenty years, and are still growing at a rate of over 5 percent a year. Exploding populations of these wild geese, particularly in the New York City area, have caused significant problems for public health officials. Large flocks of Canada Geese have been seen landing on public reservoirs and depositing their feces in great quantities, thereby causing large surges in the coliform levels of public drinking water supplies. Water department officials have been hard pressed to keep chlorine levels high enough to control all the bacteria.

In many areas, Canada Geese have been also congregating in public parks where people feed them. This situation causes the geese to interrupt their normal migration cycles and take up permanent residence in these parks. Only twenty years ago, it was rare to see a Canada Goose in a city park or on a public beach. Canada Geese were most often sighted in their familiar V-formations flying south, or were sometimes seen in cornfields, feeding on the corn left over from a recent cutting. Cornfields have always been a principal food source for Canada Geese heading south, but now cornfields are harvested with equipment so efficient that not a single ear of corn is left on the ground for geese to glean. Canada Geese have therefore been forced to turn to grass as an alternative food source.

In the New York City area, public parks have now become so severely overpopulated with Canada Geese that residents visiting those parks often have to wear boots in order to safely walk around. Goose feces is nearly ankle deep in many places, and often coats the bottoms of men's trousers, and ladies' long dresses. Park visitors can no longer allow their children to play in the grass, and people visiting cemeteries also must be very careful where they step. This threat to the public health has been totally ignored by thousands of animal lovers who enjoy seeing large flocks of geese move out of the way of their vehicles as they enter these parks.

West Nile Virus

Disease mechanisms have always been a natural part of our environment. Their presence is necessary in order to restore order when things begin to slide out of control. Disease organisms have been around for millions of years, and are a vital part of our natural ecosystem. They only rear their ugly heads when their presence is required to correct an imbalance in that system.

The recent outbreak of West Nile Virus in the US is a classic example of this natural process. The hunting of wild geese was severely restricted in the northeastern United States, mostly due to political pressure from New York City animal rights supporters. Canada Geese populations, once labeled as "threatened," soon rose to unmanageable levels. Fish and wildlife officials in the New York, New Jersey and Connecticut area found themselves under attack by animal rights supporters who considered the hunting of wild geese to be cruel and inhumane. New York City residents witnessed goose excrement growing ankle deep in their public parks, but still failed to grasp the consequences of allowing such unsanitary conditions to exist.

West Nile Virus has a very long history. When populations of wild geese began to grow out of control in Europe due to a drop in the popularity of goose hunting, many new cases of West Nile Virus were reported along goose flyways that extended from Northern Europe, down through the Middle East and into Africa.

In nature, when animal populations are allowed to increase beyond the ability of the environment to support them, disease mechanisms appear suddenly as a natural form of population control. These diseases act quickly to cull excess animal numbers and thus restore the natural balance between animal population and food supply.

Mosquitoes are still actively spreading West Nile Virus today in many parts of Europe, the Middle East, and Africa. Now that populations of wild geese in North America have also been allowed to grow out of control, the disease is appearing in the US as well. West Nile Virus has now become epidemic in the continental United States. New West Nile Virus cases are turning up all along the Atlantic goose flyway that runs from Labrador to Virginia, and are also beginning to appear along the western goose flyway that runs from Northern Canada to Mexico. This deadly disease is rapidly spreading across our nation wherever geese travel, infecting animals and humans alike.

Our desire to create a perfect world can sometimes blind us to the harsh realities of the natural world around us. We tend to ignore what we don't wish to see, and concentrate instead on what suits our pleasure. Canada Geese are beautiful birds, and the thought of a hunter shooting one may run against our grain, but a world knee-deep in goose excrement is not exactly a desirable alternative. We may wish to change the laws of nature to suit our own personal desires, but the realities of the life and death struggle going on around us will eventually come knocking at our door. God's natural world is already a perfect world. It was here long before we came, and will be here long after we're gone.

The West Nile Virus epidemic currently spreading across our nation is the direct result of our willful ignorance of the laws of nature. It appears that the efforts of animal lovers to save the Canada Goose may have actually had the opposite effect. The current overpopulation of Canada Geese in the United States is the direct result of our meddling in the laws of the natural world. The laws of nature are supreme, and we meddle with them at our peril.

Man has had some 6000 years to educate himself on the complex workings of the natural system surrounding him, but somehow in all that time, we humans have still failed to grasp the concept that we are also an integral part of that system. God's universe is a circle without beginning or end, and any alteration made to the system ultimately affects everything else in the system. We will always suffer the consequences of meddling in its operation. Once again, it seems that we've allowed the genie out of the bottle and are paying the price for our ignorance.

CHAPTER SIX:
THE ANTICHRIST

And they had a King over them, which is the angel of the bottomless pit, whose name in the Hebrew tongue is Abaddon, but in the Greek tongue hath his name Apollyon.

(Rev. 9:11)

THE HISTORY OF THE UNITED STATES has often been blessed by good luck, but there has also been an evil influence lurking as well. It behooves us therefore to also take a look at the influence this evil has had upon our past and present.

If you wake up early on a Sunday morning and turn on the radio or TV, you'll often be treated to a sermon on the coming of the antichrist. Bible evangelists have been preaching sermons about the coming of this evil figure for many decades, intimidating their listeners with stories about this frightening entity that is supposedly coming to take over the world.

Since the coming of the antichrist is a religious subject, we should probably look to the Bible to see what it has to say about this controversial figure; and it might be appropriate for us to look at the very beginning, in the book of Genesis.

The book of Genesis tells us that approximately 6000 years ago, a woman named Eve succumbed to the whisperings of the Serpent, and stole knowledge she was forbidden from possessing. The Serpent promised Eve that her stolen prize would endow her with god-like powers, enabling her to determine her own destiny. Eve, not wishing to be alone in her sin, also seduced her husband Adam into partaking of the forbidden fruit.

Ever since this ancient Bible event so aptly described in the book of Genesis, mankind has been on an eternal quest to use this stolen knowledge to create the perfect human society, a society free from all human strife and suffering.

This Genesis story actually describes the beginning of man's 6000-year journey along the path of his ultimate destiny. The Bible carefully chronicles this story of man's 6000-year quest to create the perfect human society. Many great societies have appeared on Earth over the centuries, each hoping it would be the one to finally fulfill the hopes and dreams of all humanity.

In the book of Revelation, Chapter 17, Verse 10, an angel tells us about a series of seven kingdoms that will be created by man in vain attempts to achieve this perfect society. The Bible refers to these kingdoms as beasts, because they were all destined to fail and turn on their creators, martyring millions of God's people in the process.

The exact identity, and chronology, of these seven kingdoms is therefore crucial in determining where we now stand along the path of our own destiny. The identity of these seven kingdoms, or empires, has been a source of controversy in the religious world for many decades. There has also been much confusion over the identity of the Serpent himself, who is also identified as the beast of Revelation, or the biblical antichrist.

The question is, exactly who is this antichrist, and when will he arrive? Many great minds have pondered this question down through history, and volumes of information have been written on the subject. The truth of the ancient prophecies however, was long ago sealed from the eyes of man, and therefore unavailable to us until the last days.

In the Christian Bible, you will see many different words used to describe these beasts, or antichrists. Generally a beast, in prophecy, is defined as a world-conquering empire that oppresses God's people. There have been many world conquering empires, or beasts, throughout history, but the Bible concerns itself only with those that oppressed God's people.

Words like beast, antichrist, king, kingdom, horn and crown are often used interchangeably in the Bible. The reason we must deal with so many overlapping definitions is that most ancient languages contained vocabularies of only about 5000 words. Modern languages, by comparison, often contain vocabularies of 50,000 words or more.

What this means for us, is that there may have been only one word in a particular dialect that meant nation for instance, whereas in a modern language, there might be ten times as many words available to further define that nation, such as kingdom, republic, empire, country ....well you get the idea. When early Bible writers were interpreting the Holy Scriptures, they might have used any of these words to translate that original word. Their word usage however, was generally applied with some degree of logic.

A horn for instance, is defined as a military leader or army general. A horn and a beast might be one and the same, if the horn was the leader of his empire and also the supreme commander of its armies. This was particularly true if when this leader died, his empire died with him. In this case he and his empire were considered to be one and the same entity. Such was the case with Adolph Hitler. Adolph Hitler was not only the supreme leader of the Nazi Empire, but was also supreme commander of its armies. When he died, his empire died with him.

Words like beast, king, kingdom, antichrist, horn, and crown, can often have the same meaning. These subtleties must be kept in mind when interpreting Bible prophecy. The Bible provides us with many different descriptions of these beasts, or antichrists, some of which are extremely detailed. These descriptions are repeated multiple times in different chapters and books of the Bible. This was done on purpose to provide us with multiple sources to verify and protect the truth of the Scriptures.

In the book of Revelation Chapter 17, Verses 10 and 11, an angel tells us that there will be a total of seven kings, or kingdoms, appearing on Earth to oppress God's people throughout history. The angel then tells us that five of these kingdoms, or empires, had already fallen, and that a sixth empire was just coming into existence at the time this prophecy was given, which was in the second century AD.

The angel then tells us that a seventh empire will appear sometime in the future, and that it in turn will be followed by an eighth empire, after a brief period of time. The angel also indicates that this eighth empire is somehow connected to the seventh empire before it.

History tells us that the seven great empires of the western world were all part of an unbroken chain, each empire conquering the one before it. The angel however, tells us that this eighth empire is not a part of that chain, and appears after a period of time when there "is not" an empire (Rev.17:8). This vital piece of information will prove to be very valuable when we attempt to identify this last empire to oppress God's people.

The first five beasts, or empires, to oppress God's Hebrews are well known. They are the five great empires of history, the Egyptian, Babylonian, Persian, Greek, and Roman empires. The Bible carefully chronicles the journey of the ancient Hebrews out of their oppression in Egypt, through further oppressions under the Babylonian, Persian, Greek, and Roman empires.

The prophecies of Daniel provide us with even more details about this succession of beasts, starting with Babylon, and following the trail one step further to include a sixth empire. Daniel Chapter 2 tells us about King Nebuchadnezzar's dream of a statue with a golden head, silver breast, brass belly, iron legs, and ten toes of iron and clay that will not mix together. Daniel explains to the king that this statue represents his great Babylonian Empire and the four empires that will follow it, which are in turn, the Persian, Greek, Roman, and Holy Roman empires.

In Daniel Chapter 7 we learn about a dream that Daniel has concerning four beasts. In Daniel's dream, the first beast is a winged lion that stands up as a man, the second beast is a bear raised up on one side, and the third beast is a leopard with four heads. A fourth beast is also described as a terrible creature with great iron teeth that devours everything before it. This fourth beast eventually breaks up into ten pieces, and gives rise to a "little horn" with the eyes of a man, who plucks up three of the horns, or kingdoms, before him. This "little horn" is the first of two coming antichrists

The winged lion of course, is the great symbol of Babylon. As you passed through the main gate of this ancient city, 120 of these winged lions, each about 8-feet long, stared down at you from the magnificent blue-tiled walls of Babylon's great Processional Way.

The bear raised up on one side was the symbol of the Persian Empire that consisted of both the Median, and the stronger Persian, empires.

The four-headed leopard was the symbol of Greece, divided up into four parts by the four generals of Alexander the Great, and finally, as the Iron Age arrives, we have the Iron Age kingdom of Rome that eventually breaks up into ten pieces on the old Roman peninsula.

Daniel Chapter 8 tells us about another vision that Daniel has involving a two-horned ram that is overcome by a rough goat. This he-goat then divides into four smaller kingdoms. One of these kingdoms in the latter days gives rise to a "little horn," or military leader. The angel Gabriel explains the meaning of this vision to Daniel, telling him that the ram represents the kingdom of Media-Persia, and that the he-goat represents the kingdom of Greece.

We should also note that the ram with one horn higher than the other represents the fact that the Persian Empire came up after the Median Empire, but grew to be the more powerful of the two. Both these empires came together under the emperor Cyrus, who inherited one of them from his father, King Cambyses, and the other from his uncle Darius.

The angel Gabriel also tells Daniel that the he-goat with four horns represents the kingdom of Greece, whose four generals divided up the Greek Empire amongst themselves after the untimely death of Alexander the Great, at the young age of 32. This vision provides us with an interesting detail about the little horn, or antichrist, It tells us that he also rises out of one of these four pieces of the old Greek Empire.

The sixth empire, or "Holy" Roman Empire, rose after the Roman Empire and flowered in the 16th century with the Renaissance, or "rebirth," of Rome. This sixth empire came into existence as a result of the takeover of the old Roman Empire from within, by the forces of Christianity. This sixth empire eventually split the Roman peninsula, by then called Italy, into ten pieces; five existing as strong independent kingdoms, and five existing as provinces of the church, thus forming the ten toes of King Nebuchadnezzar's dream, five strong and five weak, that would not mix together.

The identities of the seventh and eighth beasts were to be revealed only in the last days. These seventh and eighth beats are described in greater detail in Revelation Chapter 13. In Revelation Chapter 13, we read about these two beasts, or antichrists, and receive some valuable information about them. Before we review this information however, we should first review our list of empires.

MAJOR EMPIRES OF HISTORY

1. The Egyptian Empire   3400 BC

2. The Babylonian Empire   650 BC

3. The Medo-Persian Empire   550 BC

4. The Grecian Empire   330 BC

5. The Roman Empire   170 BC

6. The Holy Roman Empire   313 AD

7. unknown

8. unknown

As we mentioned previously, a beast is an empire, or antichrist, that oppresses God's people. We should also note that the beast's heads represent the nations making up the empire. In other words, if the beast has seven heads, the empire is made up of seven nations. Crowns, also sometimes called kings, represent the political leaders of those nations.

The vision described for us in Revelation Chapter 13 is of two beasts, and is given by John, who at the time is being held prisoner on the Greek island of Patmos. As John gazes out across the great Mediterranean, he sees the first beast, or antichrist, rising up out of that sea. John tells us that this first antichrist has seven heads, which means that his empire is made up of seven nations. This beast also has ten horns, or crowns, which means that three extra kings sit upon the thrones of his seven-nation empire. This is another unusual fact that may help us to identify this first antichrist.

The Antichrist

John then tells us that this first beast looks like a leopard, which backs up what the angel Gabriel told us about his Greek origins. This leopard has the feet of a bear, which means that his armies move swiftly, like the ancient Persian armies, living off the food of the land they conquer. This leopard has the mouth of a lion, which means he speaks the Babylonian philosophy that he, the emperor, is God. We then read that one of his seven heads, or nations, is put down in total defeat, but somehow miraculously recovers.

The next four verses of the prophecy tell us that this first antichrist is given power to blaspheme God, and overcome the church and its saints for a period of 42 months. We are also told that the people of the world will worship this emperor as a great figure in history because of his power and accomplishments, and that only the people of God will learn his true identity.

In Revelation Chapter 13, Verse 10, we encounter a rather unique sentence that neatly divides the descriptions of these two beasts. What this sentence does, is to describe the ultimate fate of each of these last two antichrists. The first half of this sentence tells us that the first beast, or first antichrist, causes God's people to be led into captivity, and that for this sin it is his fate to also be placed into captivity. We can skip the second half of the sentence, because it concerns only the second antichrist (the eighth beast).

We should now assemble all of the information we've gathered about this first antichrist into a list for review.

1. We know that this first antichrist comes out of one of the ten pieces of the old Roman Empire, and also out of one of the four pieces of the ancient Greek Empire.

2. We know that he plucks up three of the ten pieces of the old Roman Empire, and takes them into his new empire.

3. We learn that his empire teaches the sinful ideas of men, thereby blaspheming the laws of God.

4. We know that he rises out of the sea.

5. We know that his empire consists of 7 nations, including 3 of the 10 nations that remain of the old Roman Empire on the Italian peninsula.

6. We also know that he appoints three extra kings to rule over his 7-nation realm.

7. We are told that he looks like a leopard, or Greek.

8. We know that his armies move swiftly like the ancient Persian armies, living off the food of the land they conquer.

9. We know he speaks the Babylonian philosophy that he, the emperor, is God.

10. We learn that the army of one of his seven nations is completely destroyed, but then miraculously recovers.

11. We know that he is given the power to blaspheme God and to hold the church captive for a period of 42 months.

12. We are told that the people of the world will not view him as an antichrist, but will instead worship him as a great figure in history.

13. We learn that his ultimate destiny is to be placed into captivity.

The Antichrist

Now, armed with all this information, it should not be difficult for us to identify this first beast, or first antichrist, of Revelation Chapter 13. One detail is still missing though. We also need some sort of a timetable for his arrival. Since in Revelation Chapter 17 we learned that his empire was part of an unbroken chain of empires, we might simply look to the end of the sixth empire for his appearance. In order to do this however, we must take a closer look at the history of that sixth empire.

The sixth empire, or Holy Roman Empire, came into existence as a result of the Roman Empire being taken over from within by the forces of Christianity. History tells us that the teachings of Jesus Christ had spread throughout most of the Roman Empire by AD 100.

The pagan sun-worshippers of Rome desperately tried to prevent this internal takeover by declaring war on Christianity; but the more cruelty the Romans inflicted upon their Christian masses, the more their subjects turned toward the kindness and mercy of the Christians.

After engaging in this futile struggle with their Christian masses for many years, the Romans finally gave in and agreed to form a union with the Christian Church in order to create a new empire. This new empire, known as the "Holy" Roman Empire, was part Roman and part Christian, and governed by both a Roman Emperor and a Christian Pope. This strange anomaly lasted for over one and one half millenniums.

By the year 1800, the old Roman (now Italian) peninsula had been broken up into ten kingdoms. Five existed as independent kingdoms, and five were under church control, thus forming the ten toes of King Nebuchadnezzar's statue, five strong and five weak, that would not mix together.

The Holy Roman Empire finally came to its end when the last Holy Roman Emperor, Francis II of Austria, gave up his empire around the year 1806. The person that Francis II surrendered his empire to (and I hope you're referring to your list) was:

1. Born and raised on the Mediterranean island of Corsica, which is a part of the Republic of Genoa, one of the 10 pieces of the old Roman Empire, and also once a part of the Greek Empire.

2. He was also the man who plucked up three kingdoms on the old Roman peninsula, and took them into his new empire. The three kingdoms he usurped were the Kingdom of Naples, the Illyrian Provinces, and the Kingdom of Venice.

3. He is the man who imposed a new code of law, and a new system of weights and measures (the Metric System) upon the world, and promoted humanist philosophies, blaspheming the laws of God.

4. He rose up out of the Mediterranean Sea from the island of Corsica.

5. He led a 7-nation empire that included 3 of the 10 provinces of the old Roman Empire on the Italian peninsula.

6. He ultimately appointed three extra kings to watch over his 7 realms; they were his three brothers, Jerome, King of Westphalia, Louis, King of Holland, and Joseph, King of Spain.

7. He was a little man who stood only 5 feet 2 inches tall, and he had the black hair and olive complexion of his Greek ancestors.

8. His army was described by Tsar Alexander I of Russia, as an "army of locusts," because it lived off the food of the lands it conquered, like the ancient Persian armies.

9. He was able to seize complete military and religious control of the world, and crown himself emperor of it all.

10. The army of one of his nations, France, was completely destroyed, and its emperor sent into exile, but the army miraculously recovered when its emperor returned from exile to raise a new army.

11. On July 5th, 1809, he took the Pope of Rome prisoner, and brought him to France, thus gaining complete control of the Church and its seat in Rome for a period of 42 months. In January 1813, after a disastrous military defeat in Russia, he was forced to restore the Pope's power in return for permission to raise a new army from the ranks of the Catholics of Germany.

12. He is still admired today by the entire world as a great military leader and champion of the Enlightenment.

13. In 1821, after being held as a prisoner in exile on the island of St. Helena for many years, he died a miserable death in captivity.

Tsar Alexander I of Russia was the first person to identify Napoleon Buonaparte as the first antichrist of Revelation Chapter 13. Napoleon was indeed the man who ended the reign of the Holy Roman Empire when the last Holy Roman Emperor, Francis II of Austria, surrendered his empire to Napoleon in the year 1806. As leader of that seventh empire, Napoleon fulfilled all Bible prophecies concerning the first antichrist of Revelation Chapter 13, even adding the bloodline of the Roman Empire to his own by marrying the daughter of Francis II, and producing an heir to the throne.

When Napoleon decided to invade Russia in 1812, he came to the attention of Tsar Alexander I, a devoted student of the Bible. Alexander identified Napoleon as the first antichrist through three famous Bible references to him. The first reference was in Revelation Chapter 9, Verse 11, where he is identified as Apollyon, leader of the army of locusts. Napoleon was known to sign his name using only the letter "N." When this letter was placed before his biblical name, Apollyon, it yielded his Greek identity as "Napollyon, the Destroyer."

The second biblical reference to this antichrist was in Daniel Chapter 7, Verse 8, where he is referred to as the "little horn." Napoleon, because of his short stature and cockiness, was often referred to as the "little corporal" or "little general." In Daniel 7 he is also identified as the little horn who plucks up three pieces of the fallen Roman Empire, and Napoleon did indeed conquer three provinces on the old Roman peninsula and take them into his empire.

Alexander also read in Daniel 7, Verse 25, that Napoleon would attempt to institute new standards of "times and laws" upon the world, and noted Napoleon's attempts to institute a new International Metric System of weights and measures, and impose the Napoleonic Code of law upon Europe.

When Napoleon launched an invasion of Russia in 1812, Alexander vowed never to allow Napoleon and his "army of locusts" to invade the Russian motherland. Alexander followed biblical instructions on how to defeat this antichrist and his locust army. Alexander had read in the Bible that Napoleon's troops fed off the food of the land they conquered, like locusts, and were thus dependent upon that food to successfully complete their next flight.

Having identified this vulnerability, Alexander immediately ordered all the towns in Napoleon's path stripped of food, and burned to the ground. When Napoleon's troops arrived in each town, they found the towns totally devoid of all food and supplies. There was not even hay for their horses to eat. The horses began to eat the thatch from house roofs, and quickly bloated and died.

Napoleon began his campaign with one of the largest armies ever assembled. He left Paris with almost 600,000 troops and 10,000 horses. Napoleon's troops however were soon faced with the problem of no food to eat, and gradually began to succumb to the effects of slow starvation. The Emperor would not admit defeat however, and stubbornly drove his army on to Russia. When Napoleon crossed into Russia from Poland, an official census shows that he still had over 422,000 troops under his command.

The Russian army did not even bother to stop to do battle with the French, but simply stayed out in front of them, burning everything in their path. When Napoleon finally arrived at the gates of Moscow, his army had been reduced from 422,000 men to only 150,000 men, the same size as the Russian army. Alexander's locust-eradication tactics had been incredibly effective against Napoleon's onslaught.

The Russians finally met Napoleon in battle outside Moscow, but then retreated, allowing the French to enter the city. When Napoleon entered Moscow, he found that it too had been stripped of all food and supplies. When winter finally arrived, Tsar Alexander ordered the city of Moscow burned as well.

Napoleon and his generals were nearly killed attempting to escape the great fire of Moscow, and tried to beat a hasty retreat back to France, but things did not go well for them. Russian Cossack troops proceeded to cut the remains of the French army to pieces, and the Emperor was nearly captured. Temperatures were now hovering near 30 below zero, and Napoleon's men were dropping by the side of the road and freezing to death. When the remains of Napoleon's army finally crossed the

Neiman River back into Poland, an official census records their number at only 10,000 troops.

To save their Emperor's life, the French decided to send Napoleon on ahead in disguise. The Emperor did finally arrive back in Paris, but he was alone and defeated. His entire army had been completely destroyed by Tsar Alexander. Napoleon needed to raise a new army quickly in order to defend himself from his enemies, who were on their way to dethrone him.

Forty-two months earlier, Napoleon had taken Pope Pius XII prisoner. For 42 months, from July of 1809 to January of 1813, Napoleon held the Pope in confinement at Verona, and then at Fontainebleau. The Pope had not yet heard of Napoleon's disastrous defeat, and when the Emperor approached him with a contract granting him his freedom in exchange for the right to raise a new army from the ranks of the Catholics of Germany, the Pope gladly signed the agreement.

The following week, the pontiff found out that he'd been tricked and quickly tried to back out of the agreement by announcing that he'd made a mistake. When Napoleon heard of the Pope's announcement, he quipped with a smile, "His Holiness, being infallible, could not possibly have made a mistake."

Tsar Alexander's tactics had worked perfectly. He'd completely destroyed Napoleon's entire army of over half a million men, by using simple locust eradication tactics.

Napoleon's new army was unable to defend the Emperor for very long, and Alexander was ultimately able to march into Paris and depose Napoleon from the throne of France. The Emperor was sent into exile on the island of Elba, and it is rumored that Tsar Alexander was involved in an affair with Napoleon's empress, Josephine.

Alexander originally revealed Napoleon's identity through his name, which in the Greek language means "The Destroyer." Alexander however, failed to take note of the name of the island on which he'd exiled the Emperor, for the name Elba means "place of returning," and it was Napoleon's destiny to eventually return from Elba. Napoleon's return from exile however, was brief, and the second time the Emperor was imprisoned, it was upon the island of Helena, which in the Greek language means "Hell." Napoleon did not escape Helena alive.

Napoleon was indeed the little man who successfully ended the reign of the Holy Roman Empire. He was also emperor of the seventh empire to rule the European continent. We are told that Napoleon was defeated by the British at the Battle of Waterloo, but that is an inflated version of the truth.

In a typical battle involving Napoleon's army, it was customary for each side to experience battle losses of from 30 to 50 thousand men, or approximately 10 percent losses. If you check your history books, you'll find that at the Battle of Waterloo the British lost only about 14,000 men, making Waterloo a mere skirmish as battles with Napoleon go.

Napoleon was actually defeated long before he arrived at Waterloo. On his march to Moscow, Napoleon lost his entire army of over 500,000 men to Tsar Alexander I of Russia. It was Alexander who actually ended the reign of Napoleon, but not before the Emperor had made many changes to the world as we know it.

Tsar Alexander was the grandson of Catherine the Great, Queen of Russia. Catherine, originally from Germany, was cousin to the Queen of England, and therefore part of the royal House of David seated upon the thrones of Europe at the time. She passed on her royal heritage to her grandson Alexander. It was therefore Alexander's prophetic destiny to defeat the antichrist Napoleon. Alexander did everything in his power to save the thrones of Europe from the onslaught of the Enlightenment, but his efforts were in vain. It seems that men still wished to pursue their dream of the perfect society.

Tsar Alexander later attempted to establish religious-based agrarian societies in Russia, similar to those of the Amish people of America. These communities functioned as religious-based farming communes. They unfortunately were later replaced by atheistic socialist communes under the reign of communist revolutionaries who followed the liberal socialist doctrines of men like Karl Marx and Vladimir Lenin. These godless men had their own ideas on how human society should be run.

When these socialist revolutionaries opened Alexander's tomb in 1926 in order to desecrate it, they were met with a big surprise. His tomb was empty. You see, Alexander also knew that a second antichrist, or eighth beast, was yet to come, and planned to arrange for the demise of this final beast as well. Exactly who was this eighth beast that Alexander was so concerned about? You may find the answer to that question in another chapter of this book.

CHAPTER SEVEN:
THE METRIC SYSTEM

Thus pounds and feet and inches
Were very soon replaced
With kilograms and meters
To suit the Devil's taste.

Anonymous

HAVE YOU EVER WONDERED why we presently have two measuring systems in America? The introduction of the Metric System into the US has gone mostly unnoticed by the general public, but it may be important for Americans to take a closer look at the Metric System before simply accepting it as a replacement for our own system of weights and measures.

If you'll take a brief moment to glance at your car's speedometer, you may notice that it's graduated in both miles per hour and kilometers per hour. It wasn't until recently that this kilometers per hour addition was made underneath the miles per hour portion of the gauge. Most people regarded this change as a minor inconvenience, and really weren't very concerned about it. But nothing happens without a reason or a plan, and the plan behind the introduction of the Metric System into America needs to be examined more closely.

Up until the middle of the 20th century, the Metric System was practically unheard of in the United States. Then, in the late 1960's, foreign cars began to appear on America's highways in significant numbers as America's liberal elite began focusing their attention upon the European lifestyle. In order to repair these foreign cars, it was necessary for American automobile mechanics to use wrenches based on metric sizing. By the mid-1970's, metric road signs also started to appear on US highways, and soon, metric weights and measures were printed on goods purchased in American stores. Then, in 1975, President Gerald R. Ford officially signed the Metric Conversion Act, setting off a full-fledged invasion of the United States by the Metric System.

Most Americans had never dealt with the Metric System before, and were reluctant to have to buy a new set of wrenches just to change the carburetor on their teenager's foreign car. In fact most people in the US had no use for the Metric System, but it looked as if they were stuck with it, whether they wanted it or not. The thought that there might be any prophetic significance to the appearance of the Metric System in the United States has probably never crossed the minds of most people, but the Bible has much to say about the Metric System, and also about the person who first proposed its use.

In order for us to better understand the reasons behind the introduction of the Metric System to America, it will first be necessary for us to examine the roots of our own American system of weights and measures. To do this, we must go all the way back to the time of Abraham, the biblical progenitor of the Hebrews.

In the time of Abraham, God had already provided a set of laws and standards for His people to live by. Included in this set of standards was a system of weights and measures that was based upon a unit of measure known as the cubit. The cubit's length was based on the physical size of the Earth. The Sacred Cubit, as it was called, was a unit of measure equal to one

10,000,000[th] part of the distance between the North Pole and the center of the Earth. It was also a length equal to 25 English inches, or 7 "hands." If you do the math on this, it turns out that in the time of the book of Genesis, the Earth measured exactly half a billion inches tall from the North Pole to the South Pole. This dimension has lengthened just slightly over the millennia due to a process that scientists refer to as Post-Glacial Rebound. This process resulted in a gradual lengthening of the distance between the Earth's poles over time, due to a loss of polar mass from the melting of the Earth's polar ice caps.

Most Americans are still accustomed to using inches and feet to measure distance, and the ancient "hand" measure is still used today by people who raise horses. The height of a horse is still given in "hands" measure. The "hand" is a length of just under 4 inches (3.58" to be exact), and was originally equal to the width of a man's hand, not including the thumb. Seven of these hand measures equaled one Sacred Cubit of 25 English inches.

The ancient Hebrews used God's measuring system exclusively, and God intended for His people to use this measuring system for all time. When the ancient Hebrews scattered over the globe, they carried this cubit-based measuring system with them. The system was eventually adopted by many of the world's great nations; and, centuries later the people of Great Britain established the system all around the world in their many trading colonies. The British system of weights and measures eventually found its way into the United States by way of the English-speaking people who first founded our nation.

Our measuring system in America is still based upon this original Sacred Cubit measurement of 25 English inches, or 7 hands. Since the Sacred Cubit was an Earth-based measure, it was also used to parcel out land. The US acre is in fact a land area equal to 100 cubits, square. Since a cubit equals 25 English inches, the US acre is actually a square whose sides are each, 2500 inches long. This US acre of 2500 inches per side is still used today on practically every land deed in America, and only varies slightly from its original dimensions.

Over the centuries, this Sacred Cubit measure was corrupted in the marketplace by unscrupulous merchants, who would often try to cheat their uneducated customers. The original Sacred Cubit measure of 7 hands was eventually corrupted down to a lesser cubit of 6 hands, or 21.43 inches. This lesser cubit is still in use today in some parts of Ethiopia and Somalia. This 6-hand cubit was then further corrupted down to a 5-hand cubit of only 17.86 inches that was adopted by most other nations using the cubit measure.

The original people of God were stone masons by trade, and were often employed by other civilizations as architects and engineers, whenever large stone monuments or buildings needed to be erected. These ancient Hebrew masons were experts in mathematics and geometry, and used their cubit-based measuring system when constructing these monuments and buildings. Thus, the ancient cubit measure is preserved for all eternity in the dimensions of these ancient stone structures.

The King James edition of the English Bible tells us in the book of Isaiah, Chapter 19, Verse 19, that in the last days, one of these stone monuments would still be standing in the midst of Egypt as a testament to the Lord. And indeed, the Great Pyramid of Giza (pronounced Jeezah), the most massive stone structure ever built by man, still stands today, preserving this Sacred Cubit measurement. The Great Pyramid was designed and built by the children of Seth over 4000 years ago, and is a truly magnificent monument that contains a great wealth of information about God's universe. The Great Pyramid was specifically designed to preserve this knowledge.

The Great Pyramid is a four-sided geometric structure whose height is in the same proportion to the distance around its base, as the radius of a sphere is to its circumference. The pyramid thus preserves the value of pi, and also preserves this unique

The Metric System

geometric relationship between a circle and a square. Each side of the pyramid's base is exactly 365.242 sacred cubits in length, thus preserving the exact length of Earth's solar year. Scientists have been studying this last of the Seven Wonders for many centuries, and men like Sir Charles Piazzi Smyth, Royal Astronomer of Scotland, have been able to decipher many of its secrets.

Most people in the United States don't give much thought to what measuring system they use each day. Our English-speaking forefathers brought our measuring system with them when they first arrived in America, seeking refuge from the religious oppression of the Holy Roman Church of Europe.

The idea that the appearance of the Metric System in our nation might have any biblical significance, has probably never crossed the minds of most people. But the Bible contains a warning for us about this new measuring system that is now being proposed as a replacement for God's original cubit-based measuring system.

Since many of our goods and services now come to us from other nations, we find ourselves having to deal with the Metric System more every day. There are some people who think that it's now time for us to give up our American system of weights and measures and convert to the Metric System. These people tell us that it would be in our best interests to join in a new One World economic and social order.

This One World movement had its roots in the early 1960's on college campuses across the United States, where the idea of a religiously and socially unrestricted world was growing very popular. By the mid 1960's, liberal ideas and attitudes were popularly supported by many American university students, who drove around in little foreign cars shaped like lady bugs. Many people are unaware that these little cars were actually the brainchild of Adolph Hitler, who wanted his "people's car" to serve as the perfect commuter vehicle for the members of his new superior Aryan society. It was Hitler's intention to also create a new One World Order that would finally achieve a perfect human society for mankind. Hitler promised his many followers that he would lead them into a perfect society that would last for a thousand years.

Many liberal-thinking Americans also subscribed to this new One World theory that would eventually lead us into a more "enlightened" society. Our original Christian society, based upon Bible principles, was about to undergo a massive reformation involving the more "enlightened" New Age concepts and moral standards of Europe. As part of this reformation, it would also be necessary for us to convert our American system of weights and measures to conform to the International Metric System of Europe.

It is interesting to note that among the many Bible warnings we receive about the coming of the antichrist, is one concerning this antichrist's mission to change what early Bible writers referred to as "times and laws." Many Bible scholars now insist that this phrase should have more accurately been interpreted "weights and measures," for it was indeed the Emperor Napoleon, the man identified by Tsar Alexander I of Russia as the first biblical antichrist, who introduced us to the Metric System.

In the year 1790, Napoleon sent his personal agent, the Catholic Bishop Charles Talleyrand, on a mission to Paris to form a special committee of the Paris Academy of Sciences, in order to establish a new One World unified system of weights and measures. The purpose of this new measuring system would be to replace and entirely eliminate, the English cubit-based measuring system.

Since the cubit's length was equal to one $10,000,000^{th}$ part of the distance from the North Pole to the center of the Earth, Napoleon's new measuring unit, called the meter, was designed to be equal to one $10,000,000^{th}$ part of the distance from the North Pole to the Equator.

The Metric System

Cubit = 1/10,000,000th part N. Pole to Earth's Center
Meter = 1/10,000,000th part N. Pole to Earth's Equator

Napoleon wanted the new base distance for his meter to be measured along the Paris meridian that ran through his private estate in France known as Malmaison (Fr. evil-house). He also wanted all world standards for time, date and measure to be established along the Paris, France meridian, instead of the Greenwich, England meridian. The English however, were not about to give Napoleon that pleasure.

Since the meter's base distance was originally taken over the imperfect surface of the globe, the meter remains a flawed measure today. Napoleon named his new measuring unit from the Greek word "metron," meaning measure. The Emperor set his new Metric System into law on the 10th of December 1799, and immediately launched a program to establish this New Age measuring system all around the globe. He standardized the Metric System throughout his entire empire, which at the time included most of Europe. It was the Emperor's intention that his new measuring system would entirely eliminate God's cubit-based measuring system from the face of the Earth. Napoleon's new Metric System however, was not readily accepted by the English-speaking people of Britain and the United States. Napoleon had easily deceived the liberal French Jacobins, who overthrew their monarchy in the name of the Enlightenment. He then led them down the path to destruction, marching his armies across Europe in a campaign of terror and death unmatched in all history.

Napoleon's plans for world conquest however, were soon ended by Tsar Alexander I of Russia, a devoted student of the Bible. Alexander identified Napoleon as the first antichrist through specific references to him in the Bible. One reference, found in Daniel Chapter 7, Verse 25, mentioned that he would seek to change "times and laws." Another reference in Revelation 9, Verse 11, referred to him by his Greek name, Apollyon. Alexander also read in Daniel Chapter 7, Verse 25, that "Napollyon" would attempt to institute new standards of times and laws upon the world, and noted Napoleon's attempts to unify international standards of weights and measures and institute the Napoleonic Code upon Europe. When Napoleon attacked Russia in 1812, Tsar Alexander was ready for him. Alexander ordered everything in Napoleon's path burned to the ground. Napollyon was the biblical leader of the army of locusts, and by using locust defense tactics, Alexander was able to completely destroy Napoleon's army of more than 500,000 troops. When Alexander ordered the burning of Moscow as well, Napoleon's troops fled the city in terror and attempted to flee back to France, but were cut to pieces by Russian Cossack troops. Napoleon's army was completely destroyed by Tsar Alexander.

Alexander had completely defeated Napoleon, but the Emperor's Metric System managed to survive the collapse of the empire, and live on in the European marketplace. The same liberal-minded people who originally supported the system, still found it appealing, and did their best to carry on Napoleon's dream of a new One World economic and social order. They worked to legislate the Metric System into the laws of every nation on Earth. Even the United States government in 1975, under the administration of President Gerald R. Ford, foolishly signed the Metric Conversion Act, and moved America one step closer to the total elimination God's cubit-based measuring system.

The long-range plan of the supporters of the One World Order is to establish a single unified system of commerce and government all around the globe. In order to accomplish this task, it will first be necessary for them to create a unified One World economic, monetary, and social order, with a unified system of weights and measures. To this end, the supporters of the new One World Order have incorporated their plan into the curriculum of America's schools, and are now educating the children of America on the advantages of this new One World Order. Many Americans however, have not taken kindly to the Metric System, in spite of the barrage of propaganda in its favor. Most Americans don't particularly care for the Metric System, and have in fact refused to accept it as their national system of weights and measures.

God's holy remnant in the United States will never accept the Serpent's measuring system, in spite of the fact that it is being forced down their throats by their own leaders. The founders of this nation came here seeking freedom and independence from the moral corruption of the Holy Roman Church of Europe, which had strayed from God, and condemned Europeans to the reign of evil dictators like Napoleon and Hitler.

Our founding fathers established the United States of America as a sovereign Republic, operating "under God," where every individual was guaranteed certain God-given rights. Our founding fathers were European Protestants who rejected the corruption of the Holy Roman Church of Europe.

They established a new "free society," where the ultimate power was vested in the people themselves, and not in the government. They purposefully included the words "Creator" and "God" in the original charter they used to found this nation on July 4th, 1776, and defined the God-given rights that all free citizens would forever hold against the forces of both church and state. Now that we've successfully entered the Seventh Millennium of man's civilized existence on Earth, there is suddenly a renewed interest in Bible prophecy. Christians everywhere are re-examining the basic truths underlying the changes taking place in the world today. These truths have always been available for everyone to plainly see within the pages of the Bible.

It is important for all people to understand the subtle forces behind the social change now being proposed in American society. It is also necessary for all Americans to closely examine the motives of those who are telling us that they have only our "best interests" in mind. The founding fathers believed that a society based upon the plans and desires of sinful men, was a society doomed to certain failure, and that if men wished to establish a better society on Earth, it would need to be a society based upon the supreme laws of the Creator.

CHAPTER EIGHT:
THE YEAR WITHOUT A SUMMER

"I believe in one God, Creator of the universe. That He governs it by His providence. That He ought to be worshipped. That the most acceptable service we render Him is doing good to His other children. That the soul of man is immortal and will be treated with justice in another life, respecting its conduct in this."

Benjamin Franklin

ARE WEATHER EVENTS ACTUALLY ACTS OF GOD? In this modern world, we don't often consider weather events to be true Acts of God, even though we still refer to them as such. Many years ago however, all weather events were thought to be Acts of God. The early American colonists were firm Bible believers who attributed all natural events to the will of God.

The Bible tells us that in the last days, people will worship a God of their own making, denying His power to act through such things as weather (II Tim 3:5). The following story recounts an extremely unusual weather event that took place in early American history. Could this weather event have been a true Act of God? You decide.

It happened almost two hundred years ago, and still stands today as one of the strangest events in all of history. It has long since been forgotten by most New Englanders, but nevertheless constitutes an important happening in the early history of our nation. Its occurrence needs to be more closely examined if we are to determine its prophetic significance, if any.

It is difficult for modern Americans to think of weather events as being prophetic, but sometimes, as in the case of the biblical Flood, they can be. When the early settlers of this nation arrived on America's shores, they entered into a much harsher climate than the one they left behind in Europe. Temperatures in the New World could rise as high as 110 degrees Fahrenheit in the summer, and dip to as low as 40 below zero in the winter. These European adventurers found the harsh New England climate very difficult to deal with. European weather was somewhat similar to that of the New World, in that the four seasons of the year were well defined, but in Europe the warm breezes off the Atlantic Ocean tended to keep Europe's summers and winters much milder than those of America.

Since in those days people lived closer to, and thus were more dependent upon, the weather for their survival, the weather in any given year was of paramount importance to every New Englander. The survival of their summer crops was vital to the year's food supply, and even a sudden unexpected thunderstorm could damage crops and food storage facilities. Weather events were therefore viewed with much more concern than they are today.

In today's world, we do not tend to view the cycles of the seasons with as much trepidation as the early New Englanders, but occasionally we do experience a hurricane or other natural disaster that can threaten our homes or property.

The early settlers of New England on the other hand, saw God as an integral part of their lives, and drew a direct connection between God and all natural events, including the weather. So-called natural disasters, such as earthquakes and hurricanes, were thought to be Acts of God brought down upon mankind as vengeance for sins that men had knowingly, or unknowingly, committed. Early in the 19ᵗʰ century, a natural event occurred that was so unusual as to stand out as just such an occurrence. It happened in the year 1816, and is remembered today as the most unusual weather event in history. The year 1816 was recorded in American folklore as "The Year Without a Summer."

As with many natural disasters, the possibility that this event may have been of prophetic significance was never even considered by those who recorded it. But its story must be told, if we are to understand the role this event may have played in the overall order of our nation's prophetic destiny.

Spring of the year 1816 found many New Englanders anxiously waiting for winter to loosen its harsh grip on their land; but spring that year was unusually late, and for some unknown reason, the winter cold just refused to release its grip on the New England landscape. Deep winter snows in the Appalachian Mountains were not melting normally, and winter temperatures were not moderating as spring approached. Many people had also noticed that there was an eerie corona around the moon each night, and there was a constant haze in the heavens that just wouldn't go away. On most normal evenings, New Englanders were able to gaze up into the sky and admire the thousands of stars that dotted the heavens, but now not even a single star was visible in the evening sky.

Everyone was anxiously waiting for the summer sun to start warming the ground, signaling the start of the growing season, but somehow that warmth seemed to be avoiding New England this year.

As summer grew closer, many people were becoming concerned that something was very wrong. It wasn't that the days weren't warm, in fact some days it was downright hot. The problem was that each night the temperature would drop sharply, and by morning there would be frost on the ground. Many new England farmers were hesitant to plant their fields, only to have the frost kill the young plants, but spring was rapidly waning, and planting had to be undertaken soon or there would be no crops at all.

There were only so many weeks to the growing season in New England, and in northern New England particularly, it was necessary to take advantage of every one of them. By June 1st, most farmers had already sowed their fields. By June 5th, daytime temperatures were hovering in the 80's and it appeared as if spring had finally arrived.

Then, on June 6th, a strange thing happened. It began to snow! It started slowly at first, but by the following morning it was snowing much harder, and by that evening, all New England was in the grip of a full-blown blizzard! Four days later, New Englanders opened their doors to snow drifts over two feet deep in many places. All the way up and down the Appalachian mountain chain from Canada to Pennsylvania, it seemed the US had suffered a major snow and ice storm!

No one in the Northeast could remember anything like this ever happening before in the month of June. Even native Americans, who had lived in the New World for thousands of years, had no recollection of a similar event ever occurring. Most New Englanders were concerned and frightened by all this strange weather, and thousands of them crowded into churches seeking answers from their religious leaders. They wanted to know why God was visiting this terrible punishment upon them. What had they done to provoke His wrath? Church leaders had no answers for them; they too, were helpless to explain this act of God's vengeance.

This strange weather gave rise to all sorts of unusual events. Birds that had flown north to enjoy the warm New England summer, now suddenly found themselves in conditions for which they were totally unprepared. They desperately began to seek the warmth and refuge of people's homes and barns. If anyone were careless enough to leave a window open, a bird would immediately fly in. There was not enough room in the houses and barns of New England to fit in one more bird.

All over the Northeast, the carcasses of hundreds of thousands of birds that had died from cold and lack of food, lay strewn everywhere on the New England landscape. The same fate was also visited upon insects. Millions of them, brought out of their winter sleep by the brief warm spell, now lay dead or dying on the pure white blanket of snow. It was a hellish scene to behold, and many people thought it was the first sign of the dreaded Judgment Day. There were prophecies in the Bible about the stars disappearing from the sky, and dark days occurring just prior to the Day of Judgment. Churches all over New England were suddenly packed with people praying for forgiveness for whatever it was they had done to offend God. There was widespread fear of starvation as farmers stared out their windows at the ruined crops in their fields.

Many new spring lambs were also unable to survive the bitter cold, and quickly succumbed to the severe winter-like temperatures. Adult sheep, having just been freshly shorn, could not now defend themselves from the lethal cold, even though they sought the shelter of barns. All over the Northeast, thousands of farm animals lay dead or dying. The high temperature on many days hovered at only 30 or 40 degrees Fahrenheit.

By June 10th, ponds that had completely thawed out were once again frozen over with ice up to an inch thick. Water troughs were capped with ice every morning, and had to be emptied and refilled daily so farm animals could drink. It was June, but most people were still walking around in their winter attire.

Many families willingly sacrificed their farm animals in order to provide meat for their families and neighbors. There would be no hay or grain with which to feed the animals anyway. Most smokehouses were working overtime trying to preserve as much meat as possible, just in case the winter-like weather continued. Neighbor would help neighbor. It was the Christian thing to do.

Another strange phenomenon was also noted. In the larger river valleys, a thick fog would form each night due to the reaction of cold air and the warmer river water. Some farmers noticed that crops planted near these larger rivers were protected from the frost by this heavy warm mist. This phenomenon was well known to the ancient mountain dwelling Indian tribes of Aztecs and Incas, who purposely dug such water canals through their fields in order to generate these same protective mists. Farmers whose fields were located near these larger rivers would at least be able to provide some grain for their less fortunate neighbors.

After the first snowfall in June, the weather began to improve a bit, and it appeared as if things had finally changed for the better. People were soon joking about all the talk of impending doom the unusual weather had generated. Warm weather just before the cold had brought out all the insects, and most of them had been killed off by the frost. For the following few weeks the weather had grown much warmer, and now, with no insects to attack their crops, it appeared as if the growing season would be the best ever seen in New England.

Then, just when it seemed that all the danger had passed, the weather again turned sour. By the 4th of July, men could be seen working outside in their overcoats, at noontime, in full sunshine! It seemed as if winter had once again returned. Killing frosts were recorded on June 10th, July 9th, August 13th, and all through the summer. Snow fell again in July, and on the tops of many mountains, trees were reverting back to their fall colors.

On the top of most mountains, green leaves were now turning bright autumn colors and falling to the ground. Most New England mountain tops were totally devoid of foliage. The whole scene looked like an eerie winter landscape. Never in New England history had there been a summer like this one.

The Year Without a Summer

Women were soon forced to venture out into the forest to dig for roots, and children were sent out to pick what wild berries were to be had, either for canning, or to be used immediately as food. Milk from cows and sheep, which fed on grass, soon became a major food source for both children and adults. What extra milk there was, could be turned into cheese and stored for later use. Farmers desperately hunted squirrels, rabbits, deer, and just about any other creature with four legs that could be eaten. Freshwater fish from ponds and rivers were also used to supplement the meager food supplies.

As summer advanced, and food became even scarcer, food rationing became a normal part of life for most New Englanders. Corn, which comprised the major food crop, was over 90 percent destroyed by the weather. What corn was for sale in the marketplace, sold for 10 times normal price, and was mostly unfit for human consumption. Only the hardiest grains could be purchased in stores, and at vastly inflated prices. There was no decent food to be had anywhere for any price, and rich and poor alike suffered from the starvation rations. People began to call 1816 "The Poverty Year" and "Eighteen Hundred Froze to Death."

Decades earlier, America's great scientific genius, Benjamin Franklin, had performed studies on this kind of phenomena. Franklin wrote that this weather was often caused by layers of ash in Earth's upper atmosphere that prevented sunlight from reaching the ground. It was Franklin's opinion that such phenomena often resulted from volcanic activity elsewhere on the globe. Franklin also noted that abnormal sunspot activity sometimes contributed to the process. Franklin taught that it was often a combination of reduced sunspot activity and volcanic dust in the upper atmosphere, that resulted in reduced solar radiation here on Earth. Franklin demonstrated that a magnifying glass could not even set a piece of paper on fire under such circumstances.

Most New Englanders were totally unaware of the fact that on April 11th of the previous year, a giant volcano had indeed exploded on a remote island in the Dutch East Indies. This huge explosion sent a giant cloud of volcanic ash high up into Earth's atmosphere, much higher than normal rain clouds could reach; for it was rain that normally brought such clouds back down to Earth.

Volcanoes sometimes erupt slowly, but at other times they can erupt in one gigantic explosion. The modern Mount St. Helens volcano erupted in just such an explosion, but it luckily exploded sideways instead of straight upward, depositing very little ash into the upper atmosphere. The giant East Indies volcano called Mt. Tambora however, exploded straight upward, depositing enormous quantities of volcanic ash into Earth's stratosphere.

In 1815, Tambora's volcanic cloud created winter-like conditions in southern Europe, ruining food crops and leading to widespread starvation and food riots in France. The cloud undermined the emperor Napoleon's attempts to return from exile, and left him with no food with which to feed his army. The cloud completely destroyed the already ruined economy of France. Was this God's punishment upon Napoleon? As champion of the Enlightenment, Napoleon had left a massive trail of death and destruction across Europe in his efforts to establish his new One World order, and it seems that the fledgling United States had unwittingly supported Napoleon in these efforts.

Napoleon fought more battles, conquered more land, and killed more people than any previous world conqueror, and he did much of this with money from the United States of America.

When Napoleon needed money to finance his world conquest, he worked out a deal with the US government for the purchase of the French-owned Louisiana Territories. Napoleon's personal representative, Catholic Bishop Charles Talleyrand, was sent to work out the deal with America's politicians. Talleyrand used the money he obtained to support Napoleon's war effort in Europe. It was therefore America's money that financed more death and destruction than the world had ever seen.

American politicians however, were unconcerned with the human slaughter their money was funding in Europe. They were too busy celebrating all the newfound riches suddenly made available to them through their new Louisiana Purchase. Tambora's cloud of volcanic ash, after aiding in Napoleon's downfall, had completely circled the globe, and was now visiting its wrath upon the people of the United States.

If the church masters of New England had been more watchful over the affairs of their own government, they might have been better able to answer the questions being posed to them by their congregations.

Luckily, Tsar Alexander I of Russia had done a better job of studying his Bible, for it was Tsar Alexander who first identified the emperor Napoleon as the first antichrist. Alexander identified Napoleon by his Greek name "Napollyon," leader of the army of locusts.

When Napoleon invaded Russia in 1812, he came to the attention of Tsar Alexander, who vowed never to allow Napoleon and his "army of locusts" to invade the Russian motherland. A standard method of dealing with locusts was to fight them with fire. Farmers would often set their fields on fire to kill the locusts, and also to destroy their food supply. It took a lot of energy for the large-bodied locusts to fly on to their next destination. Burning their food was an effective way to halt their advance.

Capitalizing on this vulnerability, Alexander ordered all the towns in Napoleon's path stripped of food and burned to the ground. The tactic worked perfectly. Almost 80 percent of Napoleon's troops were lost due to starvation, on their march to Moscow. When Napoleon finally arrived in Moscow, he found that it too had been completely stripped of all food and supplies. When winter arrived, Tsar Alexander ordered the city of Moscow burned as well. Napoleon and his generals barely escaped the conflagration alive. The French army tried to beat a hasty retreat back to Paris, but without the necessary food and clothing to deal with the harsh Russian winter, most of Napoleon's troops froze to death along the side of the road in 30 below zero temperatures. Alexander's locust eradication tactics worked perfectly, he totally destroyed Napoleon's entire army of almost 600,000 men and 10,000 horses. Alexander ultimately marched into Paris and deposed Napoleon from the throne of France. If not for Tsar Alexander, the world would be a much different place today.

In Matthew Chapter 24, Verse 24, Jesus warned us about the coming of many false Christs, or antichrists. Napoleon Buonaparte was identified by Tsar Alexander as the first of these antichrists. American politicians however, identified him only as a friend. For many years in their war against the English, the Americans had allied themselves with the French. The French however, were not so much pro-American as they were anti-English. The French were easily deceived by Napoleon, and followed him down the road to ruin. If Napoleon had succeeded in his world conquest, there would be a global government in place today that would make the Roman Empire seem docile by comparison. American politicians, in their greedy quest for power and money, had been more than willing to deal with the devil.

There were some American church leaders who did object to America's purchase of the Louisiana Territories because they knew the money would be used fund massive death and destruction in Europe, but their voices went unheeded in the mad rush for land and profit in America.

All in all, the year 1816 was a time of terrible suffering for the American people. The following year marked the greatest exodus of Americans southward and westward in New England history. Many settlers, particularly those in northern sections of New England and Canada, immediately pulled up stakes and headed for warmer places. No one ever again wanted to experience a year like this one.

In retrospect, we may never know if the year 1816 was truly an Act of God or not, but it certainly was an event that changed the course of American history, and it also resulted in a terrible suffering for the American people.

CHAPTER NINE:
REVEREND MILLER'S
RAPTURE

The End in 2300 years,
The prophecy had said.
If not for Miller's Rapture,
They all would soon be dead!

Anonymous

THE FOLLOWING STORY CONCERNS one man's attempt to interpret Bible prophecy. That attempt caused quite a stir in the early church world of America. The story you are about to read took place over 150 years ago, but even today this man's work is still the subject of much discussion in religious circles. His name is Reverend William Miller, and his work still forms the basis for most modern interpretations of the book of Daniel.

In the annals of New England church history, there is no story told with more amusement than the story of Reverend Miller. Reverend Miller was one of the first persons to seriously preach the doctrine of the pre-millennium Rapture. The Rapture is still a source of great controversy in the church world today.

For those of you unacquainted with this doctrine, I should explain that it stems from a prophecy quoted in Matthew Chapter 24, Verses 40 and 41, and in Luke Chapter 17, Verses 34 through 36, where Jesus tells his disciples about an event that will occur in the last days. Jesus tells his disciples that in the last days, two men would be working in the field, and that one would be taken and the other remain; and that two women would be grinding at the mill, and one would be taken and the other left.

This prophecy was the subject of much anxious discussion in church groups that assembled every Sunday in many small New England towns and villages in the early 1800's. The prophecy was interpreted by most people to mean that the Lord himself would lift His followers up into the sky to meet Him in a great heavenly Rapture, thus saving them from the terrible tribulations of the Judgment Day. In those days there were many who believed that this great Day of Judgment was just around the corner. The early New England colonists took great interest in the Bible and all of its prophecies. Most had come to America seeking religious freedom from the oppression of the Holy Roman Church of Europe, and lived their lives according to strict Bible principles.

The Bible in Europe was for many centuries the exclusive domain of the rich and privileged classes, who could read it in its original Greek and Latin texts. The clergy of the Holy Roman Church discouraged its parishioners from reading directly from the Holy Scriptures. It was instead common practice for all interpretation of the Bible to be taken from official church doctrine. There were a few of the elite, educated in the classical languages, who could read the Scriptures for themselves, but in general it was not permitted for Catholics to read directly from the Bible.

The Holy Roman Church held absolute power over all of Europe for more than 1500 years, and totally controlled the affairs of both church and state on the European continent. Around the year 1500 however, the Church fell into a state of decay, and resorted to selling tickets into heaven, signed by the Pope himself, to anyone with enough money to purchase them. These passes into heaven, called Indulgences, guaranteed the bearer free passage into the hereafter, and forgiveness for all sins.

A German priest by the name of Martin Luther rebelled against the sinful authorities running the church at the time, and decided that it was time for the common man to be able to read the truth of the Holy Scriptures for himself. Luther therefore decided to print the Bible in the common German language so that anyone could read and understand it. Luther's break from the Church marked the beginnings of what eventually became known as the Protestant Reformation. The Holy Roman Church had lost favor with Frederick the Wise, Emperor of Germany, and so the Protestant movement soon became a very popular cause. Before long, there were Mennonites, Calvinists, and many other groups joining in the rebellion. Martin Luther was ultimately able to print and release his common language version of the Holy Bible.

John Wycliff, an Englishman, had previously issued an English version of the Holy Scriptures around the year 1400, but it was attacked by the Roman clergy and banned in England in the year 1408. Then, in 1525, an Englishman named William Tyndale traveled to Germany to visit Luther, because he wanted to utilize Luther's work to create an English version of the Bible as well as the German version. Tyndale was later convicted of heresy, and hanged.

About this time, England's King Henry VIII became embroiled in a bitter dispute with the Pope over all the women he was continuing to marry and divorce. Henry's many wives were an embarrassment to the Church. It was customary in those days for Rome to approve all royal marriages, but Henry challenged the Church's authority to overrule his decisions. This dispute ultimately resulted in a break in relations between England and Rome. Henry established the new Anglican Church of England, and now, with Rome out of the picture, an English version of the Bible was finally be printed and released. Years later, this English Bible eventually evolved into the version we know today as the King James Edition. It wasn't until later in the 17th century that this English language Bible was made commonly available to the lower classes. You can imagine everyone's excitement at finally being able to read and interpret the Holy Scriptures for themselves.

This Protestant Reformation sparked a great exodus of Christians from Europe into the New World. Tens of thousands of religious pilgrims left the shores of Europe seeking the social and religious freedoms now available to them in America. America became the New Israel (Ephriam) for Europeans with dreams of emigrating to the Promised Land. Practically everyone in America carried a Bible under one arm, and the book provided an almost exclusive source of reading material on those long winter evenings in many New England homes.

New England stood as the gateway to the Promised Land. As newcomers arrived daily from Europe, they were offered 100-acre parcels of land to homestead; all they had to do was clear the land and build a house on it. Many homesteaders would then sell their property to the next person coming down the road and needing a place to stay. They could then move on westward and begin the process all over again.

New England thus became the pathway for all hopeful new pilgrims to the New Israel. It was in this spirit of hope and promise that Reverend Miller preached his message of salvation to the world. As newcomers arrived daily to take up residence in the New World, churches sprang up in every New England town and village. Ambitious church missions were launched into foreign lands by missionaries seeking to spread the gospel to all who would listen, and new churches of every known denomination were constructed in every New England town. There were Baptists, Lutherans, Deists, and churches of every known religious doctrine. Since everyone was free to interpret the Scriptures as he or she wished, it was only natural that all these different churches would emerge. William Miller was born into this period, the son of a farmer in the northwest Massachusetts town of Pittsfield.

When Miller was 4 years old, his family moved down into New York's Hudson River Valley and settled near the town of Hampton, New York. Miller married at the age of 22 and relocated to nearby Poultney, Vermont, where he became an active member of the Deist church.

Now the Deists were a purist sect that rejected many of the Roman Catholic doctrines taught in the Christian Bible, such as the virgin birth. It wasn't that the Deists rejected the entire Bible; they just thought that it had been infected with many pagan doctrines of the Roman Church; so the Deists worshipped the purest creation of God as reflected in nature, and in God's natural world. The Deist Church was founded upon humanist philosophies.

Miller eventually went off to fight in the War of 1812, and was exposed to some of the harsher realities of life, which were to have a lasting effect upon the future direction of his career. Miller returned from the war thoroughly disillusioned with human society. He soon found his salvation in the born-again teachings of the Baptist Church. Miller became curious about the ultimate destiny of mankind, and decided to begin delving into the great mysteries of the Bible. Miller studied the Bible in great depth for over two years, and decided to undertake the task of deciphering the secret codes that provided the chronology of Bible events. The end-time prophecies of Daniel were of particular interest to Miller, who found a pattern in the numbering system used in Bible prophecy that he believed held the key to the correct calculation of the time of end. Miller's discov-

Reverend Miller's Rapture

eries were destined to make him a famous figure in the early Millennium movement in America.

Popular belief at the time held that the Bible contained secret codes that could help to reveal the timing of prophetic events. One of these codes, found in II Peter 3, Verse 8, revealed that a day to the Lord was as if a thousand years to man. Believers therefore held that man's civilized existence on Earth was limited to a seven-thousand year period, or seven God-days, the last thousand years of which would be spent under the personal dominion of the Lord himself.

Since the Adam and Eve event had been calculated by Bishop Ussher as occurring around the year 4000 BC, it was thought that the Lord's return would come at the end of the sixth day, or around the end of the second millennium AD. Miller's calculations for this Second Coming, or "Second Advent" of the Lord led him to believe that the time of the end would occur on or about the year 1843. Miller's calculations for the time of the end were based upon the 2300-year prophecy of Daniel Chapter 8, Verses 14 through 23. This prophecy (a day for a year, Num. 14:34, Ezek. 4:6) referred to a time when the last of the biblical beasts, or "transgressors," would have "come to the full."

For those of you unacquainted with the rules of prophecy, the word "beast" in prophecy refers to a world-conquering empire that oppresses God's people. The prophecies of Daniel concerned themselves with these beasts, or world empires, that would come into existence to oppress the Hebrews people.

The Bible tells us that the twelve sons of Jacob fathered the original twelve tribes of God's people. When Jacob bestowed God's special blessing upon his son Joseph, Joseph's brothers became jealous and sold Joseph into slavery in Egypt. But Joseph, blessed with the gift of prophecy, soon found favor with Pharaoh, and rose to a high position in Egypt. As God's punishment for their evil deed, Joseph's brothers and their families eventually found themselves in slavery to the Egyptians. The

Egyptians were the first in a long series of empires to oppress God's people. The Egyptian Empire was followed in turn, by the Babylonian, Persian, Greek, and Roman empires. It was mostly the prophecies of Daniel that Miller was concerned with, for the succession of these biblical beasts, or empires, was vital in determining the exact time of the end.

History tells us that around 900 BC, the Israelites decided to give up their farming and sheep herding lifestyle, and establish a new city-state in the Holy Land. This city-based society soon caused the Israelites to fall into a state of great moral decay (Judges 2:11). They turned away from God in favor of a politically ruled society. The Israelites abandoned their religious roots and became self-serving, adopting the more politically correct philosophies of their new government. They even began to war with the more orthodox southern tribes of Judah and Benjamin. This angered God greatly, and so He decided to send the children of Ishmael to avenge His anger upon them. God sent the Assyrians down out of what is now Iraq, to war against the 10 tribes of Israel and take them into captivity.

The sins of God's people continued however, and a few centuries later, even the tribes of Judah and Benjamin ultimately strayed from God's laws. So God again sent the people of Ishmael down to take these remaining two tribes into captivity. It was during this period of this Babylonian Captivity, that the prophet Daniel recorded his end-time prophecies for us.

Daniel found favor with Babylon's great King Nebuchadnezzar by accurately interpreting a dream the king had experienced. King Nebuchadnezzar's dream was of a great statue with a golden head, silver breast, brass belly, iron legs, and ten toes of iron and clay that would not mix together. Daniel told the king that this statue represented his great Babylonian Empire and the four lesser empires that would follow it, which were in turn, the Persian, Greek, Roman, and "Holy" Roman empires. Miller knew that the exact identity of these empires was vital to accurately tracking Bible chronology.

Later, during the reign of Babylon's King Belshazzar, Daniel received a vision of his own concerning four beasts. Miller taught that these four beasts also represented the same four great world empires. The four beasts are described for us in Daniel Chapter 7.

The first beast was a lion with eagle's wings. Reverend Miller identified this winged lion as the Babylonian Empire. This exact winged lion actually was the symbol that the Babylonians used to represent their empire.

The second beast of Daniel 7 was a bear raised up on one side with three ribs in its mouth. Miller identified this bear as the symbol of the Medo-Persian Empire. Miller taught that all prophecies were represented in more than one place in the Bible in order to provide the reader with verification from more than one source. He compared the bear raised up on one side in Daniel Chapter 7, to the ram of Daniel Chapter 8, whose second horn came up after the first horn, but grew to be the higher of the two. Miller taught that the Persian Empire rose after the Median Empire, but was the more powerful of the two, and was therefore represented in this fashion.

The two empires of Media and Persia are represented in prophecy as one, because both came together under the Emperor Cyrus. The three ribs in the mouth of the bear represented the Babylonian, Lydian and Egyptian kingdoms over which Cyrus ruled.

The third beast of Daniel's dream was a leopard with four wings and four heads. Miller taught that this leopard represented the Greek Empire, and that the leopard's four wings and four heads represented the four swift armies commanded by the four generals of Alexander the Great. Miller compared this leopard with four heads, to the rough goat of Daniel 8 with four horns, also representing Greece. When Alexander the Great died unexpectedly at the young age of 32, his four generals divided up the Greek empire into four parts. General Lysimachus got Thrace, Hellespont and the Bosphorus to the north, General Ptolomy received Egypt, Lydia, Arabia and Palestine to the south, General Seleucious was granted Syria and all Asian lands to the east, and General Cassander was given Macedonia and Hellas to the west.

After the three great empires of Babylon, Persia and Greece had all disappeared, the most powerful beast of all suddenly arrived on the scene. This fourth beast was dreadful, terrible in strength, and exceedingly strong. It had teeth of iron, for the Iron Age had now arrived; and with iron weapons to overcome the inferior bronze weapons of its enemies, this fourth empire was able to extend its influence over a much wider area than any of its predecessors.

Miller had no doubt that this fourth beast represented the Roman Empire. He taught that this beast's ten horns represented the ten kingdoms that would eventually arise from the breakup of the Roman Empire. Miller compared these ten horns to the ten toes of King Nebuchadnezzar's statue, five of iron, and five of clay. When we add the preceding empire of Egypt to this list of empires, we now have the biblical identity of the first six empires of the western world.

1. The Egyptian Empire   3400 BC

2. The Babylonian Empire   650 BC

3. The Medo-Persian Empire   550 BC

4. The Greek Empire   330 BC

5. The Roman Empire   170 BC

6. The Holy Roman Empire   313 AD

Miller's calculations were not to be taken lightly, Miller had access to much detailed information about Bible and church history that was in many ways superior to the information available to us today. Miller's calculation for the time of the end was based upon a time when the last of the beasts to oppress God's people will have come to its end, or when "the transgressors are come to the full." This statement, recorded in Daniel Chapter 8, Verse 23, led Miller to believe that this same date would also precede the Judgment Day.

Miller followed the trail of these world empires, or beasts, all the way from the ancient Egyptians, to the Holy Roman Empire, and was convinced that the end of the Holy Roman Church would mark the end of the world, and the Lord's return.

Through many elaborate mathematical calculations, Miller determined that the end of the last beast, or Holy Roman Empire, would be on or about the year 1843. Miller was convinced that a heavenly Rapture would occur just prior to this date.

He began to preach his Rapture theory in many local churches, and found many people receptive to it. Millennium fever was growing rapidly in the 1830's, and Miller was soon overwhelmed with invitations to speak at churches all over New England.

In 1838, Miller wrote a book entitled *"Evidence from Scripture and the History of the Second Coming of Christ."* Miller's book stirred a great deal of interest from those who wanted to believe that the Judgment Day was near.

Miller's preaching eventually came to the attention of Joshua Himes, a Baptist minister from Exeter, New Hampshire who saw great potential in Miller's message. Himes became Miller's manager, publisher and public relations director, and soon the Millerite movement was a major force in New England religion.

Himes published books, pamphlets and newspapers on Miller's predictions, and organized evangelistic crusades all over the Northeast. Miller was not an ordained minister, but had obtained a license to preach, and was soon recognized as a popular source of Bible instruction. Miller had his share of scoffers though, and was sometimes pelted with eggs at public evangelistic gatherings and camp meetings.

Miller and his followers came to believe that the true church of God would be Raptured up to heaven in advance of the Judgment Day. Miller's message appealed to the vanity of his audiences. Everyone wanted to believe that they were the exclusive members of God's elect, and would therefore be granted a special place in heaven. Miller acknowledged that his calculations might not be totally accurate, and when his original date passed with no Rapture occurring, Miller was forced to recalculate his figures.

Miller's new calculations revealed the date of the end to be the year 1845. His calculations were so precise that he announced the Rapture would occur at exactly midnight on October 22, 1844. Miller's following had now grown to over 50,000 people, and included hundreds of thousands of others who were extremely interested onlookers. Many of Miller's followers sold all their worldly goods and made final preparations to be Raptured up to heaven on the fateful day. The eyes of all New England were on the Millerites as they prepared for the great event.

On the evening of October 22nd, 1844, as midnight approached, the hilltops of New England were the scene of elaborate preparations, as Millerites, many dressed in white Ascension Robes, and others sitting in metal washtubs, patiently waited to be lifted up to heaven in the upcoming Rapture. When midnight passed with no Rapture occurring, the disappointed Millerites were forced to return to their homes in humiliation, as the townsfolk laughed and jeered at all the fools who'd fallen for Miller's great folly.

After what eventually became known as the "Great Disappointment," Miller's following rapidly fell apart, and the remaining Millerites eventually reformed under the leadership of a woman named Ellen Gould White. The Millerite movement was written off as a great failure in the annals of church history.

The Great Disappointment was viewed as just another example of the silly fanaticism of religion. It was the 19th century, and science was rapidly emerging as a much sounder way of revealing the mysteries of life, and of the universe. Religious theory was constantly being shown to be unscientific, and therefore flawed.

There is however, one very interesting fact that should be taken into account before rendering any judgments on Miller's work. Miller's calculations were based upon the end of the "last of the transgressors" occurring 2300 years after the going forth of the commandment to rebuild Jerusalem.

It has recently been discovered that the Vatican copy of the Greek Septuagint (from which our English Bibles were originally copied) records Daniel's Chapter 8 prophecy as occurring 2400 years after this event, instead of the 2300 year figure appearing in our English Bibles. It seems that an error occurred during one of the many Bible transcriptions that took place over the centuries. A pen stroke was apparently dropped from the number by mistake. When corrected for this error, Miller's date for the end of the last empire to oppress God's people becomes 1945 instead of 1845.

Interestingly, 1945 did in fact mark the end of Adolph Hitler's infamous Nazi Empire, which was indeed dedicated to the complete annihilation of the Hebrew people. Perhaps the Holy Roman Empire was not the last of the biblical beasts after all. We will further explore this complex issue of the exact identity of the biblical beasts in some later chapters of this book.

CHAPTER TEN:
HOW CHRISTMAS WON THE WAR

"It would be particularly improper to omit in this first official act, my fervent supplications to that Almighty Being who rules over the Universe, who presides in the councils of nations, and whose providential aids can supply every human defect; that His benediction may consecrate to the liberties and happiness of the People of the United States."

George Washington – Inaugural Address

AMERICA'S HISTORY HAS OFTEN BEEN FRAUGHT with peril and strife, but somehow fate has always acted at the last minute to turn the tide of battle and save the day. The following story chronicles a strange but true event in American history. It is sometimes a good thing for us to revisit history in order to gain a better understanding of the many ways that God has watched over us in difficult times.

In the latter part of the 18th century, a new spirit of independence and adventure was sweeping through the civilized world. European settlers were arriving daily upon America's shores with dreams of finding freedom and opportunity in the New World. America was rich in natural resources, and could be a "well of plenty" for people willing to work hard to make their dreams come true. These early colonists came here fleeing the class system in place in Europe allowing wealthy European nobles to economically oppress the lower classes.

When England's King George began imposing heavy tax burdens upon his American colonies, the colonists rebelled. Most of them were having a difficult enough time just trying to survive in the New World. It was nearly impossible for farmers and businessmen alike to show a reasonable profit for all their hard work and struggles in America; and worse, the colonists had no political representation in England with which to address their many grievances to the government.

This "taxation without representation" policy soon gave rise to the famous Boston Tea Party, where a group of angry colonists dumped a load of English tea into Boston Harbor in protest of the government Tea Tax. This ultimate act of defiance drew a final line between the British colonists and their Monarch. After the Boston Tea Party, the Crown proceeded to impose harsh new penalties on its American colonies. Thus the stage was set for the American Revolution.

In 1775, an armed conflict arose in Boston, Massachusetts, and the historic battles of Lexington and Concord followed Paul Revere's now famous ride. On July 4th, 1776 the American Revolution was made official with the issuance in Philadelphia of a document known as the Declaration of Independence. This founding document was hand-written by a man named Thomas Jefferson. Jefferson was the man who would ultimately be credited with the creation of a new form of government based upon the God-given rights of all free men.

From the first symbolic battles in Boston, it was soon apparent that the American Revolution was going to be a long war. The British had a well-trained army, and plenty of money with which to finance a war. The British were also more experienced in how to successfully conduct and win a conventional war.

They were also known to hire professional armies from other nations to do their fighting for them. The American colonists on the other hand, were poorly trained and poorly equipped to fight in any prolonged conflict; they did however have one thing on their side, and that was the will and determination needed to survive severe hardship.

The American Revolution has many interesting stories to tell. One of the most interesting of them all however, is the story of how the entire direction of the war was changed by the outcome of a single battle. This battle was one of the strangest battles in all of history; its memory is forever preserved for us in the famous painting of George Washington's crossing of the Delaware River. The real story of this battle however, is not often told, and the story needs to be told, because this was the single battle that changed the course of American history. It's also the story of how Christmas won the war.

That' right, if it weren't for Christmas, the United States of America might not even exist today. The American Revolution you see, was a very difficult war for the American colonists to fight. You must remember that they were fighting against the army of their own government.

There were also many colonists still loyal and sympathetic to the Crown of England; but King George had levied many severe taxes upon his colonies, and these taxes amounted to an unbearable burden for most people. A small group of American colonists therefore decided that it was now time to fight for their independence from the Monarchy. King George however, was not about to allow his colonies to separate from the mother country. Many colonists were hesitant to join in any rebellion, not because of any loyalty to England, but because they considered it foolhardy to attempt to fight against the well-trained troops of the British army. Also, many colonists feared retribution from the British if the war were lost. Anyone caught siding with the revolutionary cause could be branded a traitor and hanged. The American Revolution therefore, did not get off to a very good start. There was much sympathy for the colonial cause, but most people were convinced that this rebellion was doomed to certain failure. The few skirmishes after the 4th of July declaration only served to reinforce those fears.

England however, had one big disadvantage in this war, and that was distance. The Atlantic Ocean was a large, cold expanse that took many months to cross. It was difficult therefore, for the British to deliver large numbers of troops to put down a rebellion. European armies often numbered 500,000 men, but sending an army that size across the Atlantic was nearly impossible, and so the number of troops and supplies that could be quickly delivered to fight in this war was severely limited. It was important therefore, for the British to maintain their ties with British sympathizers in the colonies who could provide their troops with food and lodging. The British were known to burn the homes of those not supporting the Crown.

The British launched their first attack at Bunker Hill in Boston. The colonials were routed, but the battle ended up little more than a draw. Since New York City was considered to be the economic heart of America, the British decided to launch their first major assault there. A large military force was dispatched from England to carry out this mission.

America's General George Washington had very little experience fighting a European style war, and with inexperienced troops as well, didn't stand much a chance of defending Long Island against the more experienced British regulars. In August of 1776, a force of 30,000 British troops was dispatched to Long Island. They quickly routed Washington's Colonial Militia, and sent them scrambling for cover. The British had an easy time chasing the Americans off Long Island, and it was plainly evident that the Colonial Militia was no match for British regulars. General Washington found it impossible to defend New York City from the British, because the city was surrounded by water and at the mercy of British sailing vessels that could bombard it with cannon fire from all sides.

In October, the Battle of Valcour Island on Lake Champlain in upper New York State did not go much better for the Americans. The colonials gave a good account of themselves, but this battle too, was lost. Another formal battle in White Plains, New York resulted in a similar defeat for Washington. The Colonials lacked the skills and training needed to effectively fight a conventional war, so in this battle as well, they lost more ground to the British.

When the British captured Fort Lee, in Westchester, General Washington's assistant, General Lee, began competing with Washington, hoping to replace him as commanding general of the Continental Militia. Lee purposely did not move quickly enough to support Washington's army, and thus hastened Washington's retreat. Washington decided to move what was left of his army south into New Jersey to see if he could bolster its dwindling numbers with volunteer forces from the New Jersey Militia. Washington's numerous defeats however, were becoming too well known, and with the British already in New Jersey, about 2000 soldiers from the New Jersey Militia refused to reenlist.

The negative outcome of all these military battles was not doing much for American morale. Support for the war was beginning to wane, and many people were beginning to think they were correct in their original opinion about the effectiveness of the Colonial Militia. Washington's troops were now having a difficult time begging food or lodging from local farmers.

Washington now headed for Princeton with only about 3000 men from his original 20,000-man army. By December, things were not looking very good for the Colonial cause, and Washington was involved in a desperate retreat from the British. He personally led his army's rear guard, burning bridges and knocking down trees in order to block the progress of the pursuing

British army. By the time Washington reached Princeton, he had only about 400 men with him. Two thousand men from the Pennsylvania Militia joined up with Washington when they finally reached the town of Princeton.

Winter was upon them, and General Greene's troops were also headed for Princeton. General Lee was slow to respond, and delayed moving his troops. Lee's lack of action eventually resulted in his capture by the British. The remainder of Lee's army under Sullivan continued on to Washington's location. General Greene headed for New Jersey as well, and General Gates came down from Fort Ticonderoga to also join up with Washington's forces.

Washington and his harried army now headed for the safety of the Pennsylvania side of the Delaware River. The British were close behind and attempting to overtake them in order to put a quick end to the war. When the Colonial Militia finally arrived at the Delaware River, they needed boats to cross, and found them at a local business that used heavy boats to ferry its goods across the river. Washington commandeered every boat in the area, not only so he and his men could cross, but also to prevent the British, who were close on his trail, from being able to cross as well. Washington realized that the success or failure of the entire Revolution was now in his hands. His Militia was in a complete shambles, and the British were closing in for the kill.

Washington managed to successfully relocate his army to the other side of the Delaware, but his men were short on food and supplies; and without proper winter clothing, many were now freezing to death in the cold. Many Militiamen did not even own shoes, and wrapped their feet in layers of rags in order to insulate them from the snow. Morale in the colonies had sunken to a new low, and Colonial troops were no longer able to beg food or shelter from their own countrymen, who were now deathly afraid of British retaliation.

Trail of Prophecy

The British General Howe was well aware of Washington's plight, and not very concerned about Washington's army in its present state. He knew that time would only serve to make Washington's situation worse, and therefore decided to set up winter camp on the New Jersey side of the river, and wait for spring.

Washington's troops were deserting the Militia in droves and heading for home. Everyone was convinced that the Revolution was doomed. To add to Washington's plight, half of his troops' enlistment contracts would be up in a few days on January 1st. Washington needed a miracle, and he needed it fast. If he could not pull off a military victory soon, the American Revolution was over.

Washington sent word to the Continental Congress that he was badly in need of reinforcements, and also food and clothing. Unfortunately, reinforcements were not in the offing; in fact, the Continental Congress itself was now in complete disarray and worried about its own future. Washington's situation was growing more desperate by the hour, but desperate situations often test the mettle of great men, and General Washington was a man of great determination and resolve.

Most of the descriptions of Washington's winter encampment in Pennsylvania begin with the following phrase:

It was a cold, snowy Christmas Eve, but George Washington and his troops had very little to celebrate.......

Well, this is where we begin our story, because the truth is, that in 1776 Christmas was not celebrated in America; in fact, if someone were caught celebrating Christmas, they could be tarred and feathered and run out of town on a rail! There were no Christmas trees, no Christmas cards, and no Christmas gifts. Christmas was a Holy Roman holiday, and not celebrated in Protestant America.

How Christmas Won the War

That's right, the biggest holiday of the year was almost unknown in America. In fact, the people who arrived on America's shores to originally found this nation, came here specifically to get away from Christmas. They were European Protestants who hated the pagan holidays of the Holy Roman Church of Europe.

The pagan Germanic tribes of Eastern Europe had worshipped the evergreen tree for many centuries because of its ability to seemingly defy death and keep its green leaves all winter long. These pagans would go into the forest to cut down an evergreen tree, and then bring it into their homes to decorate it with silver and gold. This was in direct violation of Bible instructions given in Jeremiah Chapter 10, Verse 3. The Christians of America would never even think of celebrating such a pagan holiday as Christmas.

The British however, had hired a group of German mercenary soldiers called Hessians to help them fight the Americans, and the Hessians did celebrate all the holidays of the Holy Roman Church, especially the Christmas holiday. It was this fact that was to spell a great opportunity for the Americans.

The British General Howe had left a force of about 1500 Hessians on the eastern shore of the Delaware River to keep an eye on Washington's army. It was Christmas Day and the Hessians were busy stuffing themselves with food and drink in their usual Christmas celebration. Outside, it was snowing fiercely, and by late evening most of the Hessians were sound asleep in a drunken stupor.

Thus far, all the battles of the Revolutionary War had been fought as formal military battles, with neatly lined up rows of soldiers facing each other in traditional military style. Washington decided that it was now time for a drastic change of tactics. He decided that from now on he would fight in the style he was more accustomed to, the style of the American Indian. Washington knew that all the successful armies of history moved quickly and surprised their enemies by striking where they weren't expected, at a time when they weren't expected. So from now on Washington would utilize only light cannon and fast horses, and make surprise moves that his enemies would not expect.

While it would have been a breach of military etiquette to attack the Hessians on Christmas Day, Washington was desperate and had to make a move soon, or the American Revolution was lost. So Washington decided to attack the Hessions on the day after Christmas. This might be better anyway, because most of the Hessians would be hung over from their excess of food and drink the night before.

Washington still had approximately 6,000 troops under his command, so he devised a plan to cross the Delaware River after dark on Christmas night. He broke his army into three parts, and planned to cross the Delaware at three different locations in order to completely surround the enemy.

Under the cover of a raging blizzard, Washington attempted the hazardous crossing with his men at a point nine miles north of Trenton. Washington had commandeered all boats in the area, and was in possession of some 50-foot transport boats from a nearby factory that were capable of ferrying his horses and light cannon across the river. Washington planned to surround the Hessians at Trenton, and attack from all sides at dawn, forcing them into a quick surrender.

The river was packed with ice however, and the raging blizzard was making conditions for the crossing nearly impossible. Two out of three of Washington's regiments decided it was much too dangerous to cross. Only Washington's main force made it across the river, mostly due to Washington's blind determination. By morning, General Washington and 2400 of his troops, with 18 cannon, had successfully crossed the Delaware River. The crossing took longer than expected however, and by the time it was complete, it was too late for a predawn attack. Certainly no one could have expected that Washington's army would be able to successfully cross the Delaware River in that kind of weather.

Now, with Trenton now only nine miles to the south, his army's march to town would take only a couple of hours; but Washington had to hope he did not meet up with any Hessian patrols on the road. Blizzard conditions still prevailed, but the heavy snow was actually working in Washington's favor. The blizzard was so intense that the Hessians did not think there was any possibility Washington's Militia would move in this kind of weather. The Hessians therefore, had not even sent out their normal morning patrols.

The blinding snowstorm was covering Washington's moves as effectively as if it were night. No one was able to see anything in the distance, and even the sound of the horses hooves and cannon wheels, was being muffled by the heavy snow. The sleeping Hessians were totally unaware of the Militia's approach. Washington's troops marched into town totally unannounced, and caught the Hessians by complete surprise.

Most of the Hessian soldiers were barely able to stand, let alone fight, and the Americans quickly and easily overwhelmed them. The Hessian officers had been drinking heavily the night before, and were in even worse shape than their troops. Colonel Rall, the Hessian commander, was awakened from a drunken stupor, and had trouble getting his pants on. The Americans moved swiftly into town and broke up the Hessian formations as quickly as they formed. Many of the Hessians' muskets would not fire because of the overnight dampness, but Washington's cannon performed perfectly. The Colonials raked the streets of Trenton with close range artillery fire.

The Hessian commander, Colonel Rall, was badly wounded in the battle, and many of his officers were killed. The Hessians were now surrounded and confused, and did not know where to run.

There were a total of three Hessian regiments in town. Two of them, under Rall and Lossberg, headed to the east to retreat, but found Washington's General Greene blocking their escape route. With nowhere else to run, and guns that would not fire, they threw down their weapons and surrendered to the Colonials. The third regiment headed south out of town, but was cornered against a creek and also forced into surrender. About 600 remaining Hessian troops ran for their lives through the woods and escaped to the south out of town.

The Americans shot and killed almost 100 Hessian soldiers, and took another 900 prisoner. Not a single American soldier was lost to the enemy, although two of Washington's men had fallen by the side of the road and frozen to death on the short march to Trenton. It was the most ridiculously one-sided battle in the history of warfare.

The Hessians paid heavily for their Christmas celebration of the previous evening. The Americans were now finally able to raid the British supply depots. They now had the food, blankets, and clothing they needed to keep warm, and when word got out that Washington had captured Trenton without losing a single soldier to the enemy, people were dancing in the streets. The rag tag American Militia had finally proven that it could defeat a force of trained professional soldiers. Men were now lining up to join in the revolutionary cause, and money and supplies were now available from every source. The British had been treating American colonists badly, including even British sympathizers, and everyone was hoping against hope for just this kind of victory.

When the British General Howe heard of the Trenton rout, he was absolutely dumbfounded. He just could not believe that three hardened regiments of men who fought for a living, had been so easily defeated by a bunch of rag tag American militiamen. Howe immediately dispatched General Cornwallis to retake Trenton. Washington however, anticipated Howe's move, and also the direction from which it would come.

Washington's remaining two regiments, still on the other side of the Delaware River, assumed that Washington had also failed to cross. When they heard that Washington had captured Trenton, they were shocked, and hurriedly rushed to cross the river to meet their general. Washington knew that Cornwallis was on his way however, and immediately ordered all his troops back across the river. Cornwallis finally arrived in Trenton with a small force, but was forced to wait for reinforcements to arrive from the rear. He therefore set up camp for the night.

The British could see the American campfires burning across the river, and assumed that Washington was preparing for battle the next morning, but Washington had other plans. He slipped around the British during the night, leaving just enough men to keep the campfires going so the British would think he was still there. Washington planned to get behind the British and head for the town of Princeton to the east. The Colonials took a back road so they would not run into Cornwallis' reinforcements coming in from the east. Traveling light and fast, they planned to attack Princeton as they did Trenton, totally unexpected.

With Cornwallis still in Trenton, the Militia launched a surprise attack on Princeton and, with the Trenton victory boosting their morale, were all in good spirits. The battle went back and forth, but finally, General Washington mounted his white stallion and led a charge into the enemy lines. The sight of Washington bravely charging into battle was enough to inspire the spirit of every militiaman. Their battle spirit renewed, they quickly captured Princeton.

When news got out that Washington had captured both Trenton and Princeton, the Colonial Militia was now considered unstoppable. There was a new spirit of independence in the air, and everyone now knew that the outcome of the American Revolution was assured. Soon afterward at the Battle of Saratoga, the great British General, John Burgoyne, was defeated and captured. This was the final straw for the British. They now realized that America would finally win its War of Independence. A new nation would be formed; a nation of free people living in a free society, where the true power would be vested in the people themselves, and not in the government.

Now you know the story of how it was that Christmas won the Revolutionary War. If it hadn't been for that single Christmas celebration in 1776, there might never have been a United States of America. History unfortunately, is written by the winners, and the truth is often lost in the telling. The Battle of Trenton was for obvious reasons not the most celebrated battle of the American Revolution, but it was the most important battle just the same.

The pagan roots of the Christmas holiday are well known. It is not so very strange then that the celebration of this pagan holiday should have proven to be the undoing of the pagan forces attacking the Christians of America. The Battle of Trenton was just one more example of the many strange and unusual events that governed the destiny of the people of America.

CHAPTER ELEVEN:
THE BILL OF RIGHTS

"I know of no safe depository of the ultimate powers of society but the people themselves. And if we think them not enlightened enough to exercise their control with a wholesome discretion, the remedy is not to take it from them, but to inform their discretion by education."

Thomas Jefferson

LATELY, THERE HAS BEEN QUITE A BIT OF INTEREST in America's Bill of Rights. I therefore thought it appropriate to devote a chapter of this book to this document that defines the individual freedoms we all hold as citizens of the United States of America. The Bill of Rights is the single founding document that makes the United States truly unique from all other nations on Earth.

In 1776, a solemn man sat in a small room, composing a document that would ultimately alter the future of the world. This man's name was Thomas Jefferson, and the document he was composing was America's Declaration of Independence. This document was based upon the masonic ideals of life, liberty, and the pursuit of happiness for all mankind, and expressed the hopes and dreams of inspired men like Jefferson, John Adams and Benjamin Franklin, for a better future in a free Christian society. The Declaration of Independence served as the single founding document for the United States of America on July 4th of that same year. Our nation's Constitution, and Bill of Rights, were later additions to that original charter.

In 1776, America's colonists were suffering terribly under the oppressive dictatorship of their monarch, King George of England, who was using the British army against his own people. The normal relationship between king and subject had completely broken down, and a state of martial law existed in the American colonies. America's colonists decided that their only recourse was to revolt against their own government. They were prepared if necessary, to establish a new nation where free citizens would be guaranteed their God-given rights to life, liberty and the pursuit of happiness.

Jefferson's original founding document eventually resulted in the formation of a new nation, where every citizen would be guaranteed the rights naturally endowed to him by the Creator. Jefferson hoped that sinful men, under the inspiration of God, would be able to preserve this new Republic forever.

Every nation needs a constitution, but Jefferson wanted his nation's constitution to be different from all others. He knew that every citizen would need to be afforded constitutional protection against the kind of oppression he was now suffering at the hands of his King. In the United States of America, every citizen would be protected from government oppression by rights guaranteed him under the nation's Constitution, and these rights would never be subject to government interpretation, limitation, or suspension, under any circumstances.

Jefferson, and his good friend John Adams, decided that a Bill of Rights outlining these precious individual freedoms needed to be added to our nation's Constitution. This Bill of Rights would be written in the common language so as to avoid interpretation by government courts, and would define God-given rights held by every free citizen against the government.

The Bill of Rights

This special addendum to the nation's Constitution would be a unique undertaking, and needed to be carefully planned, and well thought out.

Just prior to the drafting of this bill of individual freedoms, Jefferson was sent to France as America's Ambassador to study the new government that the French were setting up and also to review a new French version of a citizen's bill of rights. Jefferson wanted to see if this French document could be used as a basis for America's new bill. Jefferson was much distressed however, to discover that the French had not chosen to found their nation upon faith in God, but had instead chosen to reject God entirely from their nation's founding documents.

Back in America, John Adams was involved in final preparations for drafting America's new Bill of Rights. He wrote to Jefferson in France to see what information Jefferson had gleaned from the French document. Jefferson returned a list of eleven rights that he thought should be included in America's new bill, and Adams and his colleagues used Jefferson's letter in drafting the final version of America's Bill of Rights. While preparing the document however, Adams was forced to reject one of Jefferson's items concerning protections against business monopolies, because this amendment was a government duty, not a citizen's right. Adams did not want anything to confuse the basic purpose of the document, which was to define individual freedoms held by every citizen against the federal government. Two other amendments appended to the bill by America's politicians were also rejected before the final bill was approved by all thirteen states. The final Bill of Rights contained just ten amendments, each one defining a specific right that every individual held against the federal government.

America's colonists originally came to the New World fleeing the oppression of the Holy Roman Church of Europe. The Roman Church controlled most of the continent of Europe, and wielded great power in government. The colonists believed in a firm alliance between God and State, but greatly feared any alliance between Church and State. Our nation's founders wanted to provide constitutional protections for the exercise of free thought and free speech in order to allow everyone the freedom they needed to interpret the Holy Scriptures as they wished. Jefferson wanted to build upon the laws of the Creator in order to guarantee every individual his God-given rights. He therefore purposely included the words "God" and "Creator" in his original founding document, the Declaration of Independence. Jefferson wanted to guarantee every individual's right to freely express his views and to speak the truth, even if those views conflicted with popular opinion or government law. The First Amendment of the Bill of Rights addressed this crucial issue.

1. Congress shall make no law respecting an establishment of religion, or prohibiting the free exercise thereof; or abridging the freedom of speech, or of the press; or the right of the people to peaceably assemble and to petition the government for a redress of their grievances.

The Second Amendment to the Bill of Rights guaranteed every free citizen the right to keep and bear arms. America's Revolution was still fresh in the minds of the colonists. King George had used the government militia against his own people. The founders wanted to guarantee American citizens the means to defend themselves against the government's army. An armed citizenry had proved to be the best defense against the King's institution of martial law in America, thus the Second Amendment guaranteed every individual the right to keep a soldier's weapon in his home for the purpose of serving in a freely-formed citizen's militia. The Second Amendment read as follows.

2. A well-regulated militia, being necessary to the security of a free State, the right of the people to keep and bear arms, shall not be infringed.

The Declaration of Independence listed many specific offenses committed by government troops against private citizens, including the following: "(The King) has kept among us, in times of peace, Standing Armies, without the consent of our legislatures." "(The King) has affected to render the Military independent of, and superior to the Civil Power." "For quartering large bodies of armed troops among us." "For protecting them, by mock Trial, from Punishment for any Murders which they should commit on the Inhabitants of those States." "(The King) is, at this Time, transporting large armies of foreign mercenaries to complete the works of Death, Desolation, and tyranny already begun with circumstances of Cruelty and perfidy, scarcely paralleled in the most barbarous Ages, and totally unworthy the Head of a civilized Nation."

The King had permitted government soldiers and German mercenaries to take over the homes of private citizens in peacetime, without consent of the owners, and to commit crimes against the homeowners and their families without being subject to civil trial for those crimes. The Third Amendment addressed this outrage.

3. No soldier shall, in time of peace be quartered in any house, without the consent of the owner, nor in time of war, but in a manner to be prescribed by law.

The King's soldiers had conducted warrantless searches of private homes, invaded the owners' privacy, and ransacked and seized any item of property they wished, without evidence or probable cause to believe that a crime had been committed. The Fourth Amendment set down strict rules governing government search and seizure, and the protection of individual privacy.

4. **The right of the people to be secure in their persons, houses, papers, and effects, against unreasonable searches and seizures, shall not be violated, and no warrants shall be issued, but upon probable cause, supported by oath or affirmation, and particularly describing the place to be searched, and the persons or things to be seized.**

In the King's courts, citizens could be accused of capital crimes on the word of a single complainant. Private property could be seized without warrant and sold at public auction, and confidentiality and privacy were totally ignored. Courts often tried people more than once for the same crime by altering the charges, and an accused could even be forced to testify against himself. Trials were often held repeatedly until a guilty verdict was reached. Even federal soldiers serving in the military were forced to give up their rights as free citizens, and were subjected to military trials for offenses not committed during wartime or national emergency. The government courts used any excuse to abuse the rights of free citizens. These abuses by the federal courts, including the military courts, were addressed under the Fifth Amendment.

5. **No person shall be held to answer for a capital, or otherwise infamous crime, unless on a presentment or indictment of a Grand Jury, except in cases arising in land or naval forces, or in the militia, when in actual service in time of war or public danger; nor shall any person be subject to be twice put in jeopardy of life or limb; nor shall be compelled in any criminal case to be a witness against himself, nor be deprived of life, liberty,**

or property without due process of law; nor shall private property be taken for public use without just compensation.

In the lower courts, a defendant could be imprisoned indefinitely while awaiting trial, or he could be moved to a distant location and tried by people unfamiliar with his character or reputation. Judges were allowed to conduct closed-door trials, and defendants were often unable to face their accusers. Rumor and hearsay were often introduced as evidence, and defendants could be tried without benefit of counsel. These abuses by the lower courts were addressed under the Sixth Amendment.

6. **In all criminal prosecutions, the accused shall enjoy the right to a speedy and public trial, by an impartial jury of the State and district wherein the crime shall have been committed, which district shall have been previously ascertained by law, and to be informed of the nature and cause of the accusation; to be confronted with the witnesses against him; to have compulsory process for obtaining witnesses in his favor, and to have the assistance of legal counsel for his defense.**

Since judges were often biased in favor of the government that compensated them, citizens would need to be guaranteed the right to a trial by jury, even in the lower courts. The Seventh Amendment reaffirmed the uniquely American concept that the ultimate power and authority always remained in the hands of the people. The courts were therefore granted only the power necessary for them to administer the law as written. The authority to create law was reserved to State legislatures, and ultimately to the people themselves. The individual always reserved the right to appeal to the people for justice. The Seventh Amendment reaffirms every individual's right to trial by jury.

7. In suits at common law, where the value in controversy shall exceed twenty dollars, the right to trial by jury shall be preserved, and no fact tried by jury shall otherwise be re-examined in any court of the United States, then according to the rules of common law.

In government courts, judges were known to often impose high bail or fines in order to keep defendants imprisoned. Judges would ignore prescribed punishments and invent their own, thereby subjecting individuals to injury or public ridicule. These types of judicial abuses were addressed under the Eighth Amendment.

8. Excessive bail shall not be required, nor excessive fines imposed, nor cruel and unusual punishments inflicted.

The Ninth Amendment prohibits the federal government from limiting individual rights to only those specified under the Constitution.

9. The enumeration in the Constitution, of certain rights, shall not be construed to deny or disparage others retained by the people.

The founding fathers also wished to reaffirm the superior power of the States to legislate law. Legislative authority was intended to fall first to the people, second to the States, and third to the federal government, in that order. The Tenth Amendment severely restricts the legislative powers of the federal government.

10. The powers not delegated to the United States by the Constitution, nor prohibited by it to the States, are reserved to the States respectively, or to the people.

Jefferson taught that it was the nature of all governments to eventually oppress their citizens. To help prevent this from occurring, our founding fathers established this nation as an independent and sovereign Republic, not a democracy. The Republic of the United States of America was founded upon the supreme laws of the "Creator." The words "Creator" and "God" both appear on the document used to found our great nation on July 4th, 1776. The founders were very careful not to include the word "democracy" in any of the founding documents.

They would have been appalled to witness modern-day schoolchildren being taught in America's schools that they live in a "democracy," where a majority vote may overrule even the supreme laws of God. They would have been even more shocked to learn that the United States government now employs an official Ambassador to the Vatican!

Thomas Jefferson and John Adams were the true founding fathers of our nation, and God saw fit to honor them in a very special way. He took them both from us on the same day, July 4th, 1826. This day also happens to be the 50th birthday of the nation they gave birth to with their two famous documents, the Declaration of Independence, and the Bill of Rights. It was their unique destiny to be taken up to heaven on this special day. Jefferson died at home in Monticello, Virginia, and Adams died at his home in Quincy, Massachusetts. Neither knew of the other's passing.

We owe these two great American heroes a tremendous debt indeed for the many individual rights and freedoms we all enjoy as free citizens of the United States of America.

CHAPTER TWELVE:
THE THREES

The secrets of the prophecies
Deciphered now with ease,
By seeking out the path to truth
Revealed within the threes.

Anonymous

THE SIGNIFICANCE OF NUMBERS IN PROPHECY has not always been clearly understood. In prophecy, numbers are treated as signposts that must be read and followed in order to successfully arrive at a final destination. No prophecy can be accurately interpreted unless all the signposts are read, just as it is impossible to arrive at a highway destination if you miss an exit sign. Numbers therefore, are an important consideration in the accurate interpretation of prophecy. In this short story we'll explore the significance of the number three in prophecy.

Have you ever noticed how things always seem to occur in threes? It's a strange phenomenon that's been observed by many people down through history. People often take notice of these strange phenomena that don't seem to make much sense, but happen anyway. This is probably how we came to adopt the many superstitions we're always observing, like not walking under ladders, and avoiding black cats.

Those who study prophecy have also taken notice of this strange phenomenon of things occurring in threes. The number 6 for instance, has always carried a bit of an evil connotation to it, but it develops a much more evil connotation when there are three sixes, as in the number 666.

If we search back through history, we can note an interesting historical event linked to the number 666. This event was the great Fire of London that occurred in the year 1666. This fire was one of the most profound tragedies in all of history, and the fact that it occurred in a year of three sixes, did not go unnoticed by those who study prophecy.

The fire started on a Sunday in a small bakery shop near London Bridge, and rapidly spread throughout most of the old, walled city. The fire burned on for four days, and ended up destroying approximately 80 percent of London. Historians tell us that the fire may have in fact been a blessing though, because it also helped to destroy the rat-infested slums that were contributing to a plague overspreading the city at the time.

Unfortunately, in the 17th century, no one paid much attention to sanitation, because the existence of germs and their role in disease was not understood. It would be another two hundred years before a man named Louis Pasteur would come along to discover the microbe.

The infamous Black Plague had invaded London the year before the fire, and in only a few months had claimed approximately 15 percent of the city's population. The death toll from the fire however, was drastically reduced by the plague, because many of London's citizens had already fled the city for the surrounding countryside in order to escape falling ill. The fire spread rapidly through most of London's slums, destroying many of the rodents responsible for spreading the plague. The popular children's song "London Bridge is Falling Down" commemorates both the fire and the plague. The line that reads "achoo, achoo, all fall down" refers to the fact that anyone exhibiting the deadly sneeze or cough of the plague was destined to fall down dead within a day. The Black Plague was quite deadly, and once symptoms appeared, death could follow in less than 24 hours.

Many of the children's songs and nursery rhymes of that era often carried these macabre themes. There was Jack and Jill running up the hill, with Jack falling down and breaking his crown, and of course the rockabye baby in the treetop whose cradle fell down when the tree bow broke. When you add these ominous themes to that of an old witch baking little children in her oven, it makes for a terribly frightening world for the small children of those times. If we were to examine all of the grim (pardon the pun) details contained within most fairy tales, we probably wouldn't read any of them to our children.

The phenomenon of "the threes" has been noted on many other occasions as well. If we check farther back into history, we can note another instance of the phenomenon of "the threes" that occurred in the year 1333, and it was linked to the 1666 incident. It seems that 1333 was the year that the Black Plague first made its appearance in civilized times. It was discovered in Asia that year, where it took a terrible toll on many cities in southern China and India. The Black Death had mortality rates as high as 70 percent, and was actually depopulating many large cities. It was mostly confined to cities, because that was where slums and rodent populations most contributed to its spread.

There were two distinct types of plague, and both were carried by rodents One was the bubonic plague, and the other, pneumonic plague. The bubonic plague was spread by the bite of the Oriental rat flea. The pneumonic plague was usually inhaled into the lungs while its victim was sweeping a floor covered with mouse droppings.

The Threes

Rodent populations were generally spread through goods traded between nations. The early European plague of 1333 traveled to Europe along land trade routes that ran through the Middle East. The plague arrived in Europe by way of the city of Constantinople. The 1665 plague of London however, arrived by sea. London in the mid-1600's had established itself as a major world seaport, and the city was often ravaged by plague. Usually the plague arrived by way of British sailing vessels that traveled all around the globe. Wherever rodents traveled, the plague traveled with them.

The first time the strange phenomenon of "the threes" was noted in the last millennium was in the year 1111. That year marked the crowning of King Henry V of Germany as Emperor of the Holy Roman Empire. Henry received the Crown of Heaven from the Holy Roman Church in 1111. The story of how Henry got his crown is rather interesting.

Like most kings, Henry enjoyed wielding great power over his subjects. It was customary in those days for power to be shared by kings and popes. In the 12th century, the Catholic Church had succeeded in gaining temporal powers in Europe. These temporal powers gave the Church the right to own land. German land barons were soon appointed bishops of the Roman Catholic Church, and controlled vast tracts of land in Germany. As bishops of the Church, these land barons swore allegiance to the Pope, not to the King of Germany. In this way, they could also avoid paying taxes.

Henry however, wanted to rule his entire domain, not just parts of it. He therefore resolved to reclaim the lands taken away from him by the Church. Henry first tried to speak to the Pope to see if some compromise could be worked out, but the Pope refused to even discuss the issue. Henry then decided to take his land back by force.

He invaded Italy with a large army, and forced the Pope to crown him Holy Roman Emperor. He also removed the right of the Church to own land. The German land barons immediately revolted, so Henry took the Pope and his bishops prisoner until they agreed to give him what he wanted. Henry was a rather nasty fellow, but his tactics were very effective, and he finally got his way.

This brings us to the next incidence of "the threes," which occurred in the year 1222. This was the year that the Mongols launched their invasion of Europe. The Mongols were a nomadic people who wandered the plains of northern Asia raising horses, in much the same way as they still do today. The sudden appearance on the scene of the great Mongol leader Genghis Kahn however, changed the Mongols' lives forever. The Mongols had been warring with the Chinese for many years, but were never able to defeat the Chinese army. This war had been going on for so long that it was actually responsible for the construction of the Great Wall of China.

Genghis Kahn however, was a very clever and determined man. He trained his Mongol warriors to become an efficient cavalry force that could travel light and fast, and fight effectively. After Kahn finished training his army, he was able to launch a successful invasion of China and take possession of China's great wealth and power. With all this newfound wealth, Kahn was able to launch a much larger and more widespread campaign to conquer much of the known world. Kahn had once been insulted by the leader of the Arab world, and decided to extend his empire into Arab lands first.

Kahn made good use of the technology that Chinese gunpowder afforded him, and was easily able to defeat his enemies. After successfully invading both India and the Middle East, Kahn set his sights on Europe. From the Middle East, he moved northward into Eastern Europe and was able to penetrate all the way into the Russian Ukraine.

Genghis Kahn and his Golden Hordes brought the influence of the East to the West. The Mongol language and culture were to have a lasting effect upon Europe. Even today, you can still see traces of Mongol bloodlines in the faces of the Ukrainian people. The great Genghis Kahn however, had overextended his empire, and was unable to penetrate any farther into Europe. His reign of terror ended a short time later with his death in the year 1227. Genghis Kahn's fame as a world conqueror however, was permanently sealed within the pages of history.

We then come to the year 1444. In 1444, history took another prophetic turn. For many years the Christian Crusaders had marched through the city of Constantinople, and repeatedly invaded the Holy Land. These many conflicts between Christians and Muslims had gone on for over 400 years, and the lives of many Crusaders were lost in these wars, mostly due to the terrible diseases so prevalent in the Near East. In 1444 however, the Crusades were brought to a sudden end when the Ottoman sultan, Murad II, defeated an army of Crusaders at the historic battle of Varna on the shores of the Black Sea. The Roman brand of Christianity had long maintained its hold on the East, but this historic event finally allowed the Muslims to defeat the eastern branch of the Holy Roman Church at Constantinople and reclaim the Near East for Islam.

During this campaign however, the sultan made the mistake of taking hostage the 13-year old son of the European Prince of Wallachia. The boy eventually escaped his captors and found refuge in the forests of Transylvania, but he never forgot his encounter with the Turks. Many years later, this same boy, now known as Count Dracula, crossed the Danube River and repaid his debt to the Ottomans by mercilessly slaughtering over 20,000 Turks, earning Count Dracula a reputation for evil and cruelty undiminished to this very day.

This brings us to the year 1555, when a different type of prophetic event took place. In that year, a Hebrew prophet named Nostradamus published a book of prophecy that was destined to become a source of mystery and controversy for the next 450 years. Nostradamus claimed to have received visions of future events. His book of prophecy, called the *Centuries*, consisted of 10 chapters of 100 prophecies each describing Nostradamus' brief glimpses into the future. Nostradamus' many followers claimed that he was able to transcend the bonds of time to foretell the coming of great figures like Louis Pasteur and Napoleon Buonaparte.

Many of Nostradamus' prophecies were truly amazing. One of his prophecies concerned the accomplishments of the great scientist Louis Pasteur. Nostradamus actually included Pasteur's name, and told of Pasteur's discovery of the microbe, even going so far as to detail the great scientist's ultimate fate.

Nostradamus' book fascinated men for the next 450 years, as many would try to decode its secret revelations. Nostradamus however, had cleverly disguised his prophecies in much the same way as the ancient Bible prophets. Not one of his prophecies was ever successfully interpreted before its occurrence. To this day, the prophecies of Nostradamus are still quoted every time a major world event occurs.

We now leap forward to the year 1777, which brings us into more familiar times. In the year 1777, an event occurred that was destined to change the course of American history. In that year, a battle was fought that would seal victory for America's War of Independence. Until this time, the course of the American Revolution was still in doubt, but in this decisive battle in 1777, fought in Saratoga, New York, a large force of Colonial Militia soundly defeated the British army and captured the famous British general, John Burgoyne.

Burgoyne's capture had a demoralizing effect upon the British, and marked the final turning point in America's War of Independence. Everyone now knew that victory was assured, and that there would finally be a United States of America, founded upon Christian principles of freedom for every individual. A new nation of free people, living in a free society, could now be established. It marked the first time in history that a nation's power would be vested in its people, instead of its government. The United States would stand as a bastion of freedom in a bestial world.

This then brings us to another prophetic year, 1888. This year marked the date of completion for America's famous Washington Monument. This giant Masonic obelisk stands at the very center of our nation's capital, and was erected by our founding fathers as a symbol of freedom. In 1888, the members of the Masonic Order put their finishing touches on this great masonry structure that now sits in the center of the Capital Mall in Washington, DC.

This great monument, erected as a symbol of freedom, symbolized the shedding of God's grace upon America, as now mentioned in Irving Berlin's well-known song "God Bless America." The height of this giant monument was designed to represent the number of grace. That's right, another case of the phenomenon of "the threes." The Washington Monument was built to stand exactly 555 feet tall!

All this brings us to the year 1999. Not many people can remember a prophetic event occurring in the year 1999, but there was one, and it was linked to the events of the year 1555. It seems that the great prophet Nostradamus, in his famous book on prophecy, included a prediction about an event that was to take place in the year 1999 in the month of July. This was the only prophecy in Nostradamus's entire collection that gave an exact date for the prophecy's occurrence. The prophecy became known as the "King of Terror" prophecy, because many Nostradamus "experts" had incorrectly interpreted the prophecy as describing an invasion of the Middle East by a Mongolian antichrist, setting off the final battle of Armageddon. This ominous prediction fit right in with the many other "end of the world" predictions that preceded the arrival of the year 2000. More Nostradamus books were sold during this period than at any other time in history. Nostradamus suddenly became a household word, as people rushed out to buy movies, books and videos about the great prophet's famous prediction.

When the month of July passed entirely without incident, all the Nostradamus "experts" were shocked and disappointed that nothing had occurred. They quickly came up with all sorts of excuses for why the prophecy had not come true as expected, not realizing that the prophecy actually had occurred exactly as Nostradamus predicted it would. If you'd like to find out what Nostradamus really predicted to happen in July of 1999, you can find that answer in another chapter of this book under the title "Angolmois."

Nostradamus' great "Angolmois" prophecy marked the end of a Millenium of amazing prophetic events, each occurring in a year of "the threes."

CHAPTER THIRTEEN:
THE EIGHTH BEAST

He marked the ones who bought and sold
Within the marketplace,
Then swept them off to Hecatombe
To purify his race.

Anonymous

THERE IS MUCH CONTROVERSY in the religious world today over the identity of the biblical antichrist. He has been identified as everyone from the Pope of Rome to the head of the European Common Market. He has also been identified as the beast of Revelation Chapter 13. Many religious leaders still preach on the future arrival of this greatly feared antichrist. The Bible however, tells us a different story.

The Bible tells us in Mark Chapter 13, Verse 22, that there will be more than one antichrist coming to mislead the world. The Bible speaks of a series of beasts, or kingdoms, coming to oppress God's people. Beasts in prophecy are identified as kingdoms, or empires, that oppress God's people. There have been many world empires that appeared throughout history, but the Bible concerns itself only with those that oppressed the Hebrew people.

If we are going to accurately identify these biblical beasts, we must first review their history, which is carefully chronicled for us within the pages of the Bible.

In the book of Revelation, Chapter 17, Verse 10, near the very end of the Bible, an angel tells us that there will be a total of seven kings, or kingdoms, that will come to oppress God's people. The angel says that five of these kingdoms, or beasts, had already fallen, and that a sixth was just coming into existence at the time of this prophecy, which was given around the second century AD. The angel then tells us that a seventh beast, or empire, was yet to come, and that it, in turn, would be followed by an eighth beast.

It also mentions that this seventh beast "is," then "is not," then "is" again, so its spirit apparently appears, then disappears and reappears once more. The eighth kingdom is said to be "of" the seventh kingdom preceding it. This curious relationship between the seventh and eighth beasts is also mentioned in the book of Revelation, Chapter 13, where we are told that the second antichrist, or second beast, exercises all the powers of the first beast before him.

The first five beasts, or empires, to oppress God's people are well known. They are the five great empires of history, the Egyptian, Babylonian, Persian, Greek and Roman empires. The prophecies of Daniel give us even more details concerning these beasts. Daniel Chapter 2 tells us about King Nebuchadnezzar's dream of a great statue with a golden head, silver breast, brass belly, iron legs, and ten toes of iron and clay that will not mix together. When the prophet Daniel is called upon to interpret King Nebuchadnezzar's dream, he explains to the king that this statue represents his Babylonian kingdom and the four kingdoms that will follow it, which are in turn, the Persian, Greek, Roman, and Holy Roman empires. Note the metals that are used to represent these great empires, which reflect the Golden Age, Silver Age, Bronze Age and Iron Age. When we add the preceding empire of Egypt to this list, we now have the identity of the first six beasts of the Bible.

THE WORLD'S GREAT EMPIRES

1. Egypt   3400 BC

2. Babylon   650 BC

3. Medo-Persia   550 BC

4. Greece   330 BC

5. Rome   170 BC

6. Holy Roman Empire   313 AD

Daniel Chapter 7 tells us about Daniel's dream of four beasts, a lion with wings that turns into a man, a bear raised upon one side, a leopard with four heads, and a terrible fourth beast with great iron teeth. This fourth beast in the last days gives rise to a little horn (antichrist), which plucks up three of the horns before it. A horn in prophecy is a general or commander of an army.

The winged lion of course was the symbol of the great Babylonian Empire. The bear raised up on one side represents the Kingdom of Persia, consisting of both the Empire of the Medes, and the more powerful Persian Empire. The four-headed leopard is the symbol of Greece, which is eventually split up into four pieces by the four generals of Alexander the Great; and finally, as the Iron Age arrives, we have the appearance of the Roman Empire with its terrible weapons of iron.

Daniel Chapter 8 tells us about a dream that Daniel has, of a two horned ram that is overcome by a he-goat which fathers four kingdoms, one of which gives rise in the latter days to an antichrist. The angel Gabriel explains to Daniel that the ram represents the kingdom of Media-Persia and that the he-goat represents the kingdom of Greece. We should note that the ram with one horn higher than the other compares to the bear of Daniel 7 with one side higher than the other, denoting the fact that the Persian Empire rose later than the Median Empire, but proved to be the more powerful of the two. Both these empires came together as one under the emperor Cyrus.

We should also note that the sixth empire, or "Holy" Roman Empire, was the result of an internal takeover of the Roman Empire by the forces of Christianity. This new empire, part Roman and part Christian, lasted for 1700 years and flowered in the 16th century with the Italian Renaissance, or "rebirth," of the Roman Empire. The Holy Roman Empire, by then called Italy, was eventually split up into ten pieces, five existing as independent Roman kingdoms, and five existing as Church provinces, thus forming the ten toes of iron and clay of King Nebuchadnezzar's dream, that would not mix together. The identities of the seventh and eighth beasts were not to be revealed until the last days.

These last two beasts, or antichrists, are discussed in great detail in Revelation Chapter 13. John's vision of these last two beasts was received while he was being held prisoner on the Greek Island of Patmos, in the Mediterranean Sea.

In another chapter, we explored the many details surrounding the identity of the seventh beast, or Napoleonic Empire, that conquered the Holy Roman Empire in 1806. The first seven beasts, or empires, were consecutive empires; their existence is therefore quite easy to determine by following the record of history.

We should note however, that the seventh empire differs from those before it, in that it was the empire of a man instead of a nation. Daniel 7, Verse 8, tells us that this seventh kingdom has the eyes of a man, and a mouth speaking great things, marking it as the kingdom of an antichrist. This first antichrist is also described in great detail in Revelation Chapter 13. Napoleon's empire was unique in that it was the first time in history that one man was able to take complete control of both church and state and declare himself both religious and military ruler of the world.

Napoleon's identity was first revealed by Tsar Alexander I of Russia. Tsar Alexander identified the Emperor as the first antichrist "Napollyon," leader of the army of locusts, through various references to him in the Bible. When Napoleon decided to invade Russia in 1812, Tsar Alexander vowed never to allow Napoleon and his "army of locusts" to invade the Russian motherland. Alexander followed biblical instructions on how to defeat this antichrist and his locust army. Alexander ordered all the cities and towns in Napoleon's path stripped of food, and burned to the ground. The tactic worked perfectly, and Napoleon's entire army of over a half million men was completely destroyed by the use of simple locust eradication tactics. It was Alexander who single-handedly destroyed Napoleon's army and removed the emperor from his throne. If not for Alexander, the world would be a much different place today.

Alexander was the grandson of Catherine the Great, Queen of Russia, and Catherine was cousin to the Queen of England. Catherine therefore carried the royal bloodline of the House of David, seated on the thrones of Europe at the time. She passed on her royal inheritance to her grandson Alexander. It was therefore Alexander's prophetic destiny to defeat the antichrist Napoleon. The Napoleonic Empire exactly fulfilled all the prophecies of Daniel and the book of Revelation concerning the seventh empire, or seventh beast to oppress God's people.

After the fall of Napoleon's empire, Revelation 17, Verse 8 tells us that there is a period of time when there "is not" an empire. The antichrist spirit of Napoleon disappears for a time, but then reappears in the future as the dreaded eighth beast.

This final eighth empire is that of the second antichrist, or eighth beast, the most evil of all the beasts. His empire appears near the time of the end, when there are terrible technologies available with which to wage war.

In the book of Revelation, John has great difficulty describing his visions of these marvels of technology that give this second antichrist the power to make fire come down out of the sky onto his enemies in faraway cities, and perform other miracles unheard of in John's time. We find a detailed description of this eighth beast, or second antichrist, in the second half of Revelation Chapter 13.

A unique sentence neatly divides Revelation 13's descriptions of these two antichrists. As mentioned previously, the first half of this sentence describes the fate of the first antichrist Napoleon, which was to die in captivity; and the second half of the sentence describes the fate of the second antichrist, or eighth beast. It tells us that this eighth beast causes God's Hebrews to be killed with weapons, possibly on a massive scale, and that it is therefore his fate to also be killed by such a weapon. You may notice that the weapon mentioned is a sword, which was the soldier's weapon of John's time. But this weapon might be a gun if this second antichrist were to appear in the 20th century, or maybe even a phaser, if he were to appear in the 23rd century. In any case, the prophecy tells us that since he lives by the phaser, he dies by the phaser. And thus we know that this second antichrist causes God's people to be killed with weapons, possibly on a massive scale, and that it is therefore his fate to also be killed by such a weapon.

The next verse of Revelation 13 tells us that this second antichrist does not rise up from the sea like the first beast before him, but instead rises out of the Earth. The principal land mass populated by God's people after the fall of the Napoleonic empire, was Europe. We can therefore surmise that this second antichrist, or eighth beast, rises out of the continent of Europe. John then tells us that this second antichrist has two heads, and two horns.

For those of you unacquainted with the rules of prophecy, heads represent nations, and horns represent military leaders. This means that this eighth beast's empire consists of two nations led by two military leaders. John says that the two horns of this beast are like those of a lamb, symbolizing the fact that this antichrist represents himself as a messiah, coming to lead his promised people into a thousand years of peace. The prophecy then goes on to say that he speaks with the mouth of a dragon, which means he deceives his followers by tempting them with lies they will want to hear.

The next three verses were very confusing to ancient Bible scholars, because these early scribes had no understanding of the terrible technologies John was attempting to describe in these verses. The first technology this beast possessed, was the power to make fire come down out of the sky onto his enemies in faraway cities. This was John's attempt to describe the aerial bombs and rockets he saw in his vision, that were the product of technologies not understood in the second century AD. John next talks about seeing an image having the power of speech. An image, or idol, in Bible prophecy, is a creation or invention of the hands of man. John says that this invention has the power of speech, and that anyone refusing to listen to it, can be delivered up to the beast and sent off to certain death. John also tells us that this second antichrist numbers the people who buy and sell in the marketplace, and that he places these number characters onto their forearms or heads.

The last thing John tells us is that this beast's number is 666, and that it is the number of a man. All numbers in prophecy have meanings; 777 for instance is God's number, 555 is the number of grace, 999 is the number of Judgment, and 666 is the number of sinful men. Every number has significance. Since this beast's number is 666, and is a man's number, and since it is repeated three times for emphasis, we know that we are dealing with a male antichrist. This is another reference to the fact that the seventh and eighth beasts, or empires, were not the empires of nations, but were instead the empires of men. Both the seventh and eighth empires carried this distinction, since each was led by a male antichrist.

Now let us list all the information we learned about this dreaded second antichrist who leads the eighth and last empire to oppress God's people.

1. We know that his empire is not a part of the chain of the world's first seven empires.

2. We know that his empire utilizes modern technologies with which to wage war.

3. We know that he deceives his followers by telling them they are his chosen people and that he will lead them into a thousand years of peace.

4. We know that he causes God's Hebrews to be shot, possibly on a massive scale.

5. We know that it is his ultimate destiny to be killed by a gun or other hand-held weapon.

6. We know that he rises out of the continent of Europe

7. We know that his empire consists of two nations, led by two military leaders.

The Eighth Beast

8. We are told that he spreads his lies through a man-made device that has the power of speech.

9. We know that anyone who refuses to listen to this invention may be carried off to certain death.

10. We know that he numbers the people who buy and sell in the marketplace by placing numbers on the skin of their arms or heads.

Have you figured out who this second antichrist is yet? If not, you might simply walk up to any stranger on the street and ask him to name the most evil man to ever walk the Earth. The answer you receive will often be the name of this infamous second antichrist.

History records for us the fact that it was Tsar Alexander of Russia who identified the first antichrist; but what is not so well known, is that Tsar Alexander was also aware of the identity of the second antichrist as well. When Alexander died, his coffin was placed in a tomb in Moscow. In 1926, a group of Russian revolutionaries opened Alexander's tomb in order to desecrate it. When these evil socialists opened the tomb, they were met with a big surprise. His tomb was empty! Alexander was aware of many events scheduled to occur in the future, and knew of the coming of a second antichrist, who would try to complete the work of the first antichrist before him. Alexander wanted to protect the Russian people from this eighth beast by leaving them all the information they would need to defeat this most evil of all the beasts. The secret of what became of Tsar Alexander can provide us with another clue as to the identity of this last beast.

In a previous chapter we explored the works of the Reverend William Miller. Miller's work in deciphering the prophecies of Daniel centered on the date when "the transgressors are come to the full," or the date marking the end of the last beast. This biblical reference, contained in Daniel Chapter 8, Verse 23, led Reverend Miller to believe that the date for the demise of this last beast was the year 1845. We also revealed that an error had occurred in the transcription of the King James Bible as it was being copied from the text of the original Greek Septuagint. This error involved the number of years between Daniel's prophecy and the death of this last beast. The number quoted in our King James Bibles is 2300 years (Dan. 8:14), but the number quoted in the Vatican copy of the Greek Septuagint, from which our English Bibles were copied, is 2400 years. When corrected for this error, Miller's 1845 date for the death of this eighth beast, or second antichrist, becomes 1945, and that was indeed the date that marked the demise of Adolph Hitler and his Nazi Empire. Hitler's empire committed the last and greatest oppressions against the Hebrew people.

There are some very interesting facts that should also be taken into account concerning Adolph Hitler's empire, if we are to determine its significance in prophecy. Not only did it come at the exact time predicted by Daniel's prophecy, but it also fulfilled the false Messiah prophecy of a man coming to lead his chosen people into a thousand-year empire. Hitler conducted his evil campaign under the sign of the cross. The famous Nazi Iron Cross was proudly worn by many German soldiers, and the crooked cross, or Nazi swastika (Hellas cross), became infamous as the official emblem of the Nazi regime. Hitler promised his followers that he was going to lead them into a thousand-year Reich, or empire.

Hitler's family tree is very important to the revelation of his biblical identity. Some of our earliest records trace Hitler's family history from his paternal grandmother, Maria Anna Shicklgruber, who worked as a maid for the Frankelberger family of Graz, Austria. Maria, who was unmarried, became pregnant while working for the Frankelbergers. Her illegitimate son, Alois, was destined to one day become Adolph Hitler's father.

The Frankelbergers, a wealthy Jewish family, reportedly made regular payments to Maria for support of her son Alois until his 15th birthday.

Adolph Hitler was born on April 20th, 1889 at approximately 6:30 PM in the town of Braun am Inn in northwestern Austria, near the German border. As a child, Adolph attended a religious school operated by Benedictine monks, and headed by an abbot named Theodoric Von Hagen. The Von Hagen family crest, a hagenkreuz, or "hooked cross," was emblazoned on the walls of the monastery. Hitler dreamed of one day becoming an abbot himself, and adopted the hagenkreuz, or "swastika" as we now know it, as his personal symbol.

As a young man, Hitler lived in Bohemia, often associating with the rich Jews of that region. He admired their ability to make money and live an idealistic lifestyle. Hitler dreamed of one day becoming a great artist or architect himself, and designing great statues or tall buildings. Hitler's talents however, were not very impressive, and he failed miserably in these endeavors. Once, he briefly left Germany to visit his brother in Liverpool, England, but for the most part, Hitler spent all of his life in either Germany or Austria.

When World War I broke out, Hitler joined the German army as a dispatcher, marking the beginning of his 27-year war with the world. He eventually worked his way into the German Workers Party as an undercover agent, and greatly enjoyed working as a political activist. Hitler was a liberal idealist, defending the common German worker and dreaming of a greater destiny for Germany. Hitler eventually grew to hate all Jews and Catholics, and all others who rejected him. He blamed the Jews for manipulating the stock market and destroying the German economy. With so many Germans now out of work, it was not very difficult for Hitler to find sympathy for his views. Hitler was easily able to influence the common German worker with his dreams for a glorious rebirth of the German Republic.

All names have meanings, and they reveal much about the people who carry them. The name Adolph Hitler (actually Heidler) in the German language, means "pagan wolf-leader"; and Hitler was indeed a clever leader of his Nazi pack of wolves. Hitler's headquarters in Poland was known as the "Wolf's Lair," and his eastern headquarters in the Russian Ukraine was nick-named "Werewolf," The site of the KDF manufacturing plant for Hitler's "peoples'car" is "Wolfsburg," and Hitler's submarine "wolf packs" successfully terrorized the world's oceans for the entire Second World War.

Hitler carefully studied the empire of his predecessor Napoleon, and thought that perhaps Napoleon's blood ran in his veins, since the Emperor had passed through that part of Austria a century earlier, and was known to entertain young German maidens every evening in his tent. Hitler thought it was his manifest destiny to complete Napoleon's task of creating a unified One World Order, in order to enlighten modern society and rid the world of the genetically inferior races. Hitler would succeed where Napoleon failed; he would successfully invade Russia in the winter. Hitler launched his Russian invasion on the same day as Napoleon, but 129 years later, and followed Napoleon's exact route to Moscow. Hitler's Russian campaign was code-named Barbarossa, after the great German emperor Frederick Barbarossa, who had once led Germany to greatness.

Tsar Alexander however, was waiting for him. Alexander had not died and been buried in a tomb in Moscow after all, but had instead become a monk in the service of the Russian Orthodox Church. Alexander spent the last few years of his life in the service of God, and left careful instructions for Russian Church hierarchy on what to do when the inevitable invasion of the second antichrist finally came. When Hitler launched his attack on Russia, the Russians were ready for him. Russian leaders instructed their people to destroy and burn everything in Hitler's path, just as they had done 129 years earlier when Napoleon had attacked. Every home, field and forest was set on fire, and every factory and power-generating dam was blown up or destroyed. Nothing was left for Hitler's "locust army" to use. This "scorched earth" policy worked perfectly for a second time, and Hitler's army was halted in its tracks due to severe shortages of food and supplies. The Russian front eventually proved to be Hitler's undoing. He'd captured over 500,000 square miles of Russian territory and received absolutely nothing for his efforts.

With God's help, Alexander was able to posthumously defeat the second antichrist, and save the world from a second horrible fate. Daniel's prophecy for the time when "the transgressors are come to the full," described in Daniel 8, Verse 23, was fulfilled right on schedule in 1945, just as the Reverend William Miller had calculated. The eighth and final beast met his biblical fate of being killed with a hand held weapon, just as the apostle John had prophesied. Tsar Alexander had seen to it that the era of the beasts was finally brought to an end.

The events foretold in the book of Revelation seem to be slowly coming to fruition exactly as predicted within the pages of the Bible. What new events are yet to occur, still remains a mystery. The intricate movement of the great timeclock of the universe however, still continues to tick away the minutes of man's brief time here on Earth.

CHAPTER FOURTEEN: DISEASE

"Even fleas have bugs that bite 'em...
On and on, ad infinitum..."

Sir Isaac Newton – on the design of the universe

WHAT IS DISEASE and where does it come from? For thousands of years man has been attempting to solve the riddle of disease. Disease has always been a hidden enemy that strikes without warning, wreaking havoc on our everyday lives. For almost all of man's history here on Earth, the underlying cause of disease remained a mystery. It was not until the latter part of the nineteenth century that a French chemist named Louis Pasteur finally revealed the long-hidden world of the microbe.

Pasteur discovered that disease was caused by tiny living organisms, too small to be seen by the naked eye. By the time of Pasteur's famous discovery, the world's diseases had already killed and maimed more human beings than all the wars of history. Throughout the entire period of the European Crusades, more Crusaders lost their lives to disease than to battle. When the Christian Crusaders left Europe's pristine forests and ventured into the hot, dry climate of the Middle Eastern deserts, they were quickly overwhelmed by the many deadly diseases re-

Disease

siding in the polluted water wells of that region. It seems that man has always been helpless to defend himself against the ravages of disease.

For centuries, the world's largest cities were a breeding ground for all sorts of terrible diseases. In the 14th century, the infamous Black Death wiped out almost one quarter of the population of Europe. In most cities, the plague claimed the life of nearly 70 percent of its victims. This was mainly due to the fact that most city dwellers were already weakened due to poor diet and repeated attacks of cholera. No one could explain why the plague seemed to be confined mostly to cities, and particularly to the poorer areas of those cities.

Europe's largest cities drew many people from the surrounding countryside with the promise of instant riches. Those who ventured into these cities however, usually found themselves living in slums, where clean water and sanitation were practically non-existent. Most large cities lacked adequate provisions for clean drinking water, or sanitary disposal of sewage.

Rats quickly overran slum areas, feeding off the refuse from human society. When rats overpopulated an area, nature's natural population control mechanisms were soon activated to bring a plague upon the rat population, and thus reduce its numbers.

Rat plague, also known as bubonic plague, was efficiently spread throughout the rat population by a tiny flea whose technical name was Xenopsylla Cheopis. This tiny flea could also transfer the plague to human beings. During the 14th century, the bubonic plague killed almost 25 million Europeans. The First World War by comparison, accounted for only 9 million deaths.

The infamous bubonic plague was first reported in China in the year 1333. The disease attacked the human lymph system, causing large black and blue marks to appear under the armpits of its victims. This condition gave rise to the popular term "boo boo," still used today to describe a black and blue spot.

The plague normally appeared in one of two forms, rat plague and mouse plague. Rat plague, or bubonic plague, attacked the lymph system, and mouse plague, or pneumonic plague, attacked the lungs. Both forms of this disease were caused by rodents, but bubonic plague generally entered its victim's bloodstream through the bite of a flea, while pneumonic plague was generally inhaled into the lungs, usually while the victim was sweeping a floor covered with mouse droppings.

Pneumonic plague was the deadlier of the two varieties and could kill within 24 hours of the onset of symptoms. It still occurs today in the United States in our Desert Southwest, where it has now been renamed "Hanta" virus. Pneumonic plague kills so quickly that there is not enough time for antibiotic treatments to take effect.

It was most likely this pneumonic form of the disease that was responsible for the 1665 plague of London. Pneumonic plague was also known as the "Black Death" because the bodies of its victims turned blue from lack of oxygen. The London plague gave rise to the popular children's song "London Bridge is Falling Down." The line "achoo, achoo, all fall down" refers to the fact that anyone having the deadly cough or sneeze of the pneumonic plague was destined to fall down dead within the day. The great Fire of London that occurred in 1666, also contributed to this song's famous title.

The bubonic plague still survives today in many parts of Asia, Africa and the Americas. It also still occasionally occurs in the United States. About a dozen cases of the plague are reported each year in the US. These cases do not come from overseas, but occur naturally right here in our nation, in areas where sanitary conditions have been allowed to slip.

We are sometimes told that these deadly diseases arrived here from some distant country, but this is not always the case. We experience many incidents of rare and deadly diseases occurring spontaneously, right here in the United States every year.

Disease

We were told in 1999 that West Nile Virus arrived in the United States from Africa, but this type of encephalitis is actually quite common right here in the US. Many cases of infectious encephalitis occur every year within our nation's borders.

West Nile Virus is known to be hosted by wild geese, and can become epidemic in locations where waterfowl populations have been allowed to grow out of control. When this type of encephalitis is diagnosed in America's lower Midwest, it's called "St. Louis Encephalitis," and cases occurring in the upper Midwest are labeled "La Crosse Encephalitis." If a case is found in the northeastern United States, it is generally diagnosed as "Eastern Equine Encephalitis." The chief reason that this eastern strain received its own label was due to the fact that state laboratories closely monitor the horse racing industry, regularly testing the blood and urine of racehorses to screen them for illegal drugs.

When a few expensive racehorses in the northeastern US died unexpectedly of encephalitis, the disease was given the name Eastern Equine Encephalitis. Believe it or not, this American strain of encephalitis can often be deadlier than the African strains.

When the first few samples of West Nile Virus in the US were sent to national laboratories for testing, they came back testing positive for Eastern Equine Encephalitis. Routine laboratory diagnoses are often inaccurate. Nationally we've experienced many situations where sick patients at various hospitals were diagnosed with a wide variety of illnesses, only to find out later that all these victims had attended the same business conference, and were actually victims of a mass food poisoning incident. Medicine is not an exact science, and our medical professionals still have much to learn about the pathology of disease.

Louis Pasteur's discovery of the microbe caused the medical profession to reject the concept of the "spontaneous generation" of disease. It was the 19th century, and that knowledge stolen by

Eve so long ago was finally beginning to bear fruit. The miracles of God were slowly being replaced by the miracles of modern science.

It was never Pasteur's intent to destroy this "spontaneous creation" theory; he merely wished to further define it. Pasteur demonstrated that diseases spontaneously appeared in response to the sudden creation of conditions conducive to their growth. Pasteur taught that disease organisms naturally existed everywhere in our environment, and that there were reasons why they would suddenly begin to breed out of control. Pasteur showed that disease was caused by large quantities of microbes infecting a host animal, and establishing living colonies within the body of that host.

Pasteur's discovery allowed doctors to make significant improvements in the control and treatment of many diseases. Doctors soon began to associate poor sanitation with the development and spread of disease. Government officials were soon charged with the responsibility of isolating disease outbreaks and preventing their spread.

Many diseases were eventually controlled through the use of improved sanitation, medical quarantine, and the use of antibiotic drugs or chemicals. Recently however, some diseases we thought we had long ago eliminated, have been staging an amazing comeback. It seems that these diseases possess the unique ability to develop resistance to the drugs and chemicals we once used against them.

Due to their extremely short life cycles, disease organisms are quickly able to adapt to radical changes in their environment. When a disease is attacked with an antibiotic, many disease organisms are killed off; but a few always manage to survive. These survivors pass on their genetic resistance to their offspring, thereby creating new, drug-resistant strains.

The dreaded disease tuberculosis is presently staging a dramatic comeback in America's prisons. The new tuberculosis strains that are now appearing, have proven to be highly resistant to the antibiotics we once used to successfully control them. Drug-resistant strains of many other diseases as well, are now frustrating the best efforts of medical professionals to control them. Many former "miracle" drugs have now been rendered completely useless in the treatment of disease, and in fact only serve to make disease organisms stronger.

This situation occurred because medical professionals failed to recognize the true nature of disease itself. In the natural world, disease organisms serve a very important purpose. When animal populations begin to grow out of control, it is disease that steps in to correct the situation. Disease organisms kill off excess animal numbers, thereby helping to restore the balance between the animal population and its supportive environment. Disease thus serves as a useful tool to help keep order in God's complex natural ecosystem.

As animal populations are allowed to increase in size, the opportunity for disease increases proportionately. Diseases have also been known to jump the species barrier and infect other species. In today's world, our medical professionals often have to deal with diseases that jump the species barrier in this manner. Tularemia, or "rabbit fever" for instance, can jump the species barrier and infect people who handle the remains of diseased rabbits.

When any disease strikes, it initially kills off many members of the particular species it attacks. At first glance, this appears to be a horrible disaster, but in the long run the disease ultimately brings the animal species back into balance with its supportive environment. Disease therefore, is not always our enemy, but often acts as a friend.

Diseases can easily be avoided by obeying the laws of God's natural order. We can illustrate this with a hypothetical situation. If we were to send a group of people into the Rocky Mountains to live in a community completely isolated from the outside world, we should expect them to remain relatively healthy and disease free. There would be no reason for us to expect them to contract diseases from other humans, due to their extreme isolation.

If however, these people suddenly decided to dig a new drinking water well within 10 feet of their septic system, it is likely that an epidemic of cholera would soon ensue. The cholera would not come from any outside source; it would instead be generated quite naturally, in response to the sudden creation of conditions conducive to. its growth. Cholera, like any other organism, merely responds to changes in its environment; in other words, the cholera appears as soon as the conditions suitable to its growth appear. Disease organisms always activate when conditions conducive to their growth are suddenly created.

Many times in history, man has unwittingly created the conditions suitable for the development of deadly disease. In the mid-1800's for instance, cholera was a common disease in the city of London, England. A physician named John Snow decided to examine the pathology of this deadly disease.

Snow was a London anesthesiologist who had a talent for gathering and studying data. He began keeping detailed documentation on all the cholera cases reported in London by the many physicians he worked with. When Snow examined the information he'd collected, he was surprised to find out that most of the cholera cases were confined to very specific areas of the city.

Snow laid out a map of London and marked the locations of cholera outbreaks on the map. To his surprise, he found that the cholera cases were distributed over the city in the exact same pattern as the public drinking water system. Sections of the city that received their drinking water from the Thames River for instance, would all suffer cholera outbreaks at the same time; and when a case appeared in a section of the city served by a

Disease

public well, everybody using that well would soon develop cholera.

The medical community was already aware of the fact that cholera could be transmitted from person to person through contaminated food and clothing, but this water-borne link had never been previously demonstrated.

Snow soon determined that London's drinking water intakes on the Thames River were located too close to the sea. Due to the influence of ocean tides, the river water flowed backwards twice each day, allowing raw sewage dumped into the river downstream, to back up and be sucked into the city's clean drinking water intakes upstream.

Snow's pioneering work resulted in the Lambeth Company of London digging new sanitary water wells on the outskirts of London, and installing new piping into the city. Areas served by these new clean drinking water wells had their incidence of cholera reduced by over 90 percent when compared to the rest of the city. It was finally clearly demonstrated that the primary source of cholera in London was contaminated water being drawn from the Thames River.

The work of medical pioneers like John Snow and Louis Pasteur did much to contribute to our understanding of the pathology of disease. In spite of their discoveries however, many modern medical professionals still fail to recognize the true nature of disease itself.

Since disease was always considered to be man's enemy, it was always man's goal to eliminate all disease from the environment. This was a rather lofty goal, but many great strides were made in that direction. For a time it even appeared as if man might actually eradicate disease from the world.

But man's efforts to eliminate disease from his environment turned out to be no more successful than his efforts to eliminate insects from the environment. This was due to the fact that man failed to recognize disease as an essential and vital part of God's natural system. Disease plays an important role in the complex system of checks and balances that sustain life on our planet. Disease is but one of many mechanisms that exist within nature for the purpose of correcting unnatural conditions and behaviors.

In reality, disease organisms are always present in the natural ecosystem, and only present themselves as disease when the delicate balance of that system is disturbed. Diseases will then naturally activate in order to restore order to the system. When conditions suitable for the growth of a particular disease organism are suddenly created, the organism responds and produces the conditions we recognize as disease.

Disease organisms are universal. Bubonic plague, for instance, occurs right here in the United States each year in areas where rodents have been allowed to overpopulate. In short, disease epidemics are not merely accidental occurrences; they are instead the result of human ignorance.

It seems that humans still continue to bring disease upon themselves through their ignorance of natural law. When we fail to observe the laws of God's delicately balanced order, we always pay a price for our ignorance. Science is only just now beginning to reveal the true complexity of the universal system in which we all must live.

CHAPTER FIFTEEN: LYME DISEASE

The only laws that are truly just, are the laws of nature.

Anonymous

IT SEEMS THAT AMERICA has recently been bombarded with a rash of new diseases. These diseases have been popping up everywhere, threatening the health of Americans and their families. Our medical professionals are at a loss to explain where all these new diseases are coming from, and why they are appearing so suddenly.

Due to the many medical advances of recent years, most Americans now feel that they've gained some degree of control over their personal health. This has unfortunately caused many of us to lose sight of the fact that the natural world around us is a complex system with intricate mechanisms for perpetuating its existence. In this story we'll examine the origins of a newly discovered disease that is still affecting the health of many Americans today.

In the late 1960's, an economic boom in the New York City area resulted in the sudden appearance of many new suburban communities in the nearby state of Connecticut. This occurred as urban commuters from New York City moved into this rural, farming state. These upper class New Yorkers had very little exposure to the agricultural lifestyle of Connecticut locals. This sudden influx of New Yorkers brought with it many economic and social changes for Connecticut farmers and their natural ecosystem.

The farmers of Connecticut had always held full control over their state's natural environment, but as more and more New Yorkers moved in, many farms were forced to close down, and that control was lost. Native white tail deer populations soon began to increase dramatically in size, and the New Yorkers were treated to the sight of wild deer feeding in their backyards every morning. It was a novelty for them to have their city friends visit and see these wild deer feeding under the bird feeder. Many people also began to feed the deer in order to keep them coming close to their homes. It was soon a common sight to see deer feeding by the side of Connecticut's highways.

Unfortunately, many of these deer suffered from distended abdomens, due to the large quantities of grass they were now forced to consume, due to a shortage of natural forest foods. Their wild food supplies were now depleted, due to severe overpopulation and overgrazing by the large herd. Without farmers around to cull excess deer populations, there was nothing to prevent Connecticut's deer herd from growing totally out of control.

By the mid-1970's, Connecticut Fish and Game officials, charged with the responsibility of managing deer numbers, tried to encourage deer hunting in the state in a desperate attempt to reduce white tail populations. The liberal New Yorkers however, wanted nothing to do with deer hunting. The sight of a dead deer in the back of a pickup truck with its tongue hanging out, was more than most of them could stand. Connecticut wildlife officials soon found themselves under attack by groups of New York City animal rights supporters, who considered deer hunting to be cruel and inhumane.

Lyme Disease

Deer populations continued to expand, and soon, deer fences over 7 feet high had to be erected around most homes to prevent wild deer from devouring the shrubbery. Soon, deer droppings were nearly ankle deep on the ground. The health of Connecticut's deer herd began to decline rapidly as well intentioned, but misinformed, New York City animal lovers attempted to impose their brand of logic upon God's natural world.

Then, in 1975, an unusual outbreak of juvenile rheumatoid arthritis was recorded in the small town of Lyme, Connecticut. Local doctors documented over 50 cases of this debilitating disease in a small three-town area of the state. Since these cases were so unusually concentrated, health officials decided to conduct an investigation to see if some common denominator could be found for the outbreak.

The first thing Connecticut medical officials noticed was that most of these cases seemed to be occurring in the summer when children were playing outside. They also noticed that many of the cases were occurring in homes located near wooded areas. When doctors ran tests on the children to screen them for a common bacterium, they uncovered an unidentified spirochete bacterium in all test samples taken.

When the children's parents were questioned about any unusual symptoms that had occurred just prior to the appearance of the disease, they found that most of the children had exhibited a rash just prior to the onset of symptoms. The parents were then asked to bring their children in immediately if they exhibited such a rash. It was soon determined that in all cases, the children had been bitten by a tick just prior to the development of the disease, and that a rash had formed, in a bulls-eye pattern, around the site of the tick bite. Apparently the children were being infected with this crippling disease by ticks. The specific tick identified as the culprit, was the eastern deer tick, a tick hosted by white-tailed deer.

Diseases have always been a part of the natural environment. Their presence is necessary in order to restore order to the natural system if things begin to slide out of control. When animal populations are allowed to grow beyond the ability of the environment to support them, disease mechanisms will naturally activate to cull excess animal numbers, and thus restore a balance to the system. Disease organisms have always been a vital part of our natural environment. They only rear their ugly heads when their presence is required to correct an imbalance in that system. They therefore serve as a useful tool for maintaining balance and order within the system.

The 1975 outbreak of Lyme disease in Connecticut was a typical example of this natural process. The tick responsible for the outbreak of juvenile rheumatoid arthritis in Lyme, Connecticut was Ixodes Scapularis, a tick commonly hosted by white-tailed deer. As white tail populations were allowed to grow out of control in southwestern Connecticut, nature's population control mechanisms were activated to bring a plague upon the deer herd, and thus reduce its numbers. This plague was very efficiently spread throughout the herd by the common deer tick.

White tail deer have been native to Connecticut woodlands for centuries, and served as fine table fare for American colonists for over two centuries. They are still harvested for food today by modern deer hunters, although not in sufficient numbers. The early New England colonists obeyed biblical instruction that animals with cloven hooves and chewing the cud were good for food.

Unfortunately, some of these early colonists were also guilty of abusing God's natural system. In the 19th century, greedy upper class land barons cleared almost every acre of land in New England in order to sell the land for profit to immigrants pouring in from Europe. At one time, it was possible to travel all the way from Boston to New York City without ever encountering a decent patch of woods. The infinite, and uninterrupted pattern of stone walls that still covers New England today, stands in stark testimony to this massive deforestation of the New England landscape by people more respectful of personal profit than of the environment. These rich business tycoons left almost no natural habitat to support wild creatures, and as a result, many native birds, plants, and wild animals were all but eliminated from the New England ecosystem.

It wasn't until the early 1900's, when farming was replaced by industry, that our woodlands began to recover from this terrible deforestation. White tail deer suffered massive destruction of their habitat in the 18th and 19th centuries, and now are suffering once again due to the meddling of upper class New York City animal lovers, who possess university degrees, but little common sense. Man, it seems, is still blind to the intricate workings of God's natural world, and still continues to attempt to impose his will upon it. It is now estimated that a case of Lyme disease exists in every third household in some areas of suburban New York and New Jersey.

Animal rights supporters have suggested feeding the deer food laced with birth control pills, but that solution is a purely non-selective method of population control, and does not allow for survival of the smartest and fittest animals, as does the more natural solution of predation. Some have also suggested introducing wolves to control deer populations. Wolves however, much prefer to devour children and pets, as they are much tastier and easier to catch.

Most animal lovers consider the predation of the hunter to be cruel and inhumane, but I wish they could experience the screams of a young deer fawn having its intestines ripped out by a pack of coy dogs while it's still alive and its mother looks on helplessly. They would then have something to compare to the more responsible predation of the hunter, who kills quickly and cleanly with his gun, taking reasonable care that animals do not suffer unduly in the process. Hunters by the way, do not kill deer fawns.

The animal lovers of New York City, through their bold and irresponsible actions, have now brought a terrible disease upon themselves and also upon the deer they originally sought to protect. This plague of Lyme disease is now spreading across America, causing untold suffering for humans and animals alike. This degenerative, crippling disease slowly damages the body's organs over time, resulting in slow suffering and premature death for most of its victims. There are now over 17,000 new cases of Lyme disease reported in humans each year in the United States.

Man has a long history of attempting to impose his will upon God's natural order, and in every case he has paid a terrible price for his meddling.

CHAPTER SIXTEEN:
MAD COW DISEASE

In the fading light I could barely make out the profile of the large, hulking bovine standing over the carcass of the sheep it had just killed. Its cold, black eyes were glaring madly at me, as it ripped into the bloody intestines of the dead animal. I knew instinctively what to do, "Mad Cow" I yelled, "Ma-a-a-ad Cow"!

Anonymous

IN SOME PREVIOUS CHAPTERS, we learned how unnatural conditions, such as overpopulation and poor sanitation, can lead to the development and spread of disease. But can unnatural behavior also lead to disease? It would seem that cows devouring sheep would be most unnatural behavior for cattle, but who knows what goes on out in those dark pastures after nightfall.

In this short chapter we'll take a closer look at another new disease that has been making the headlines recently. It's the disease known as BSE, or Bovine Spongiform Encephalopathy, also popularly known as "Mad Cow" disease. Mad Cow disease is one of a larger group of diseases known as TSE's, or Transmissible Spongiform Encephalopathies.

TSE's are a type of degenerative brain disease that slowly turns brain tissue into a soft, spongy mass. This disease gradually eats away at brain tissue, creating thousands of tiny holes, causing the brain to take on the appearance of a sponge, hence the name "Spongiform."

This disease was first observed in the mid-1700's in sheep, and was called "scrapie" because of the strange behavior of affected sheep, who would scrape their bodies on rocks or fence posts in mad attempts to rub off their wool. These sheep eventually lost the ability to stand or walk, and would die within a few months of the onset of symptoms. This disease however, did not seem to be transmissible to human beings.

Then, in the early 1920's, two German neurologists, Dr. Hans Gerhard Creutzfeldt, and Dr. Alphonse Maria Jakob, described a form of TSE disease occurring in human beings. This form of TSE was so rare that it had only been observed in about out of every one million people. A few cases of the disease had also been documented in certain cannibalistic tribes living in New Guinea, who had the unfortunate habit of eating the brains of their deceased relatives in an act of religious symbolism. This particular disease was known as "Cannibalism Disease," or "Kuru."

Then, in 1985, a cow in Britain was diagnosed with Mad Cow disease. This cow had apparently ingested the remains of dead sheep that had been mixed in with its animal feed. The feed was manufactured through a process that involved exposing the feed to high temperatures in order to neutralize any bacteria that might be present in the feed mixture.

Since the 1950's, meat-rendering companies had been collecting bones and other meat by-products from local butcher shops and passing them on to food processors to be ground, cooked and dried into dry dog food and other animal feeds. This process was big business in the newly emerging economy of the second half of the twentieth century. No significant problems with these dry animal feeds were observed, except for a few minor cases of food poisoning due to toxins generated by certain molds and yeasts in feed allowed to become too moist.

Feed manufacturers soon began experimenting with adding fiber and filler to their feeds, such as wheat or soy protein. Many people were beginning to become concerned however, about what sorts of products were being added into animal feed, since this industry was relatively unregulated when compared to the human food processing industry.

When the first few cases of Mad Cow disease were diagnosed in Britain, it was revealed that sheep by-products had been utilized in the manufacture of the feed being used to raise these animals. This fact led to the immediate institution of a ban on the use of ruminant by-products in animal feed. TSE disease had also been observed in people who'd eaten beef products contaminated with BSE. Mad Cow disease soon made the front pages of newspapers all around the world, sparking numerous investigations into the pathology of TSE.

TSE disease had appeared in sheep as Scrapie, in cattle as BSE, and in deer and elk as CWD or "Chronic Wasting Disease." In humans it was called "Creutzfeldt-Jakob Disease" or "Kuru." TSE's had also been found in ranch-raised mink.

In 1985, Dr. Stanley Prusiner, of the University of California San Francisco, accurately diagnosed the disease through the presence of certain deformed proteins in brain tissues that he called "prions." These deformed proteins could be found in the tissues of the brain, spinal cord, eyes, tonsils, and lower intestines of infected animals. The disease was known to cause a slow, progressive degeneration of brain tissue, resulting in confusion, loss of muscle control, drooling, prostration, and eventual death for its victims.

One way this disease could be transferred from one host to another, was through the ingestion of infected tissue from dead animals. In one experiment conducted in England, brain matter from Scrapie-infected sheep was injected into the brains of cattle. Eighteen months later the cattle developed a form of BSE.

Spongiform Encephalopathy was also recorded in some people who'd received Human Growth Hormone injections from material collected from the brains of human cadavers, and also in persons who'd received eyeball transplants from cadavers.

After the disease was first identified, and a number of cases of BSE were confirmed in cattle from the UK. A ban was finally enacted (1988) on the use of ruminant proteins in the preparation of animal feed. Almost immediately, more animal-feed bans, and bans on the importation of live animals and/or animal by-products, were enacted by many other nations around the world.

A total of 150,000 cases of BSE were eventually identified worldwide, with many of those cases being traceable to the UK. A cattle feed ban in the United States in August of 1997, banned the feeding of cattle to cattle, sheep, or goats. Cattle could still be used as food by certain other animals such as chickens or pigs. And a ban was also considered for certain cattle killing systems that involved mechanically striking the brain.

A cow slaughtered on December 9, 2003 in the State of Washington, tested positive for Mad Cow disease. This was the first case of Madcow disease in the US. Unfortunately, meat from this animal had already found its way into the US food processing system. This particular cow was born in Canada prior to the 1997 feed ban, and apparently originally came from the province of Alberta, the same place that produced the only Canadian cow found to be infected with BSE. BSE develops very slowly, and its symptoms often do not appear until 4 or 5 years after original exposure.

On Dec 30, 2003, the US government decided to ban the use of "downer cattle" (cattle unable to stand) from the processing of human food. Also banned were the use of brain, spinal cord, and certain other bovine tissues for human food.

Mad Cow Disease

The modern process of manufacturing animal feed from meat by-products, includes a sterilization process capable of destroying most viruses and bacteria that might be present in the feed mixture. This sterilization procedure usually includes high temperature boiling or steaming of the feed mixture. This process was known to destroy any disease-causing microbes present in the meat by-products.

There was a slight problem however when it came to the prions involved in BSE. It was soon determined through scientific experimentation that BSE prions were able to survive this high temperature process intact. Experiments demonstrated that prions were able to successfully endure temperatures as high as 600 degrees Farenheit, thereby allowing them to survive normal sanitation procedures.

In addition to infecting sheep and cattle, TSE's were also known to infect other ruminants as well, such as wild deer and elk. In the State of Wisconsin in the year 2000, an investigation was conducted into the operation of 550 captive game farms in that state. Many of these game farms used commercial animal feeds to raise their animals, and a few cases of CWD were confirmed in some of these game animals.

This Wisconsin investigation revealed that one farmer had lost an undetermined number of deer into the wild from his CWD-infected herd. Several other farmers reported the escape of captive deer or elk that were never recovered. Two other farmers even admitted to game wardens that they had released their entire herds into the wild.

A researcher with the National Institutes of Health announced that "Given these reports of escapes and releases of game farm deer, and given the findings of CWD infected animals on Wisconsin game farms, it is highly likely that these game farm animals are the source of CWD in wild deer."

Trail of Prophecy

Does unnatural behavior (i.e. cannibalism) lead to the development and spread of deadly diseases? Only time and more investigation will provide us with the answer to this question.

CHAPTER SEVENTEEN:
LOUIS PASTEUR

"The more I study nature, the more I am amazed at the work of the Creator."

Louis Pasteur

THE PROPHET NOSTRADAMUS recorded many prophecies for us on future events. One of his most famous prophecies concerned the accomplishments of the great scientist Louis Pasteur. Pasteur's discovery of the microbe was one of the most important events in all of medical history.

Nostradamus recorded his prophecies in the form of four-line poems known as quatrains. In Chapter 1, Quatrain 25 of his famous book, the *Centuries*, Nostradamus not only included Pasteur's name, but told of the great scientist's revelation of the microbe, and also described the persecution and ridicule Pasteur suffered at the hands of his peers in the European medical community. What is so incredible about this famous quatrain is that it was recorded hundreds of years before Pasteur was born.

Not many people are familiar with the predictions of the 16th century Hebrew prophet, Nostradamus. Nostradamus was a physician and prophet who claimed to have received visions of future events. He recorded his visions in the form of four-line poems, and published them in a book called the *Centuries*. The book contains 10 chapters of 100 quatrains each. Chapter 7 was never finished, and contains only 42 quatrains.

Nostradamus' poems are not recorded in any understandable chronological order, so dating them is very difficult. Nostradamus also purposely disguised many of his quatrains in order to confuse those attempting to interpret them before their fulfillment. His quatrains are hand-written in Old French, and thus very difficult to translate. Nostradamus was also known to anagram certain words or names in order to disguise them, and often included a "punch" line, or keyword, in each of the quatrains, which when correctly translated, revealed the true meaning of the poem. Through these clever deceptions, Nostradamus was successfully able to keep his prophecies veiled for many centuries.

The prophecies of Nostradamus generated enormous controversy through the centuries, concerning his alleged ability to see through time. The world's foremost scientists and astrophysicists all agree that it's impossible for man to pierce the bonds of time to foretell future events.

Nostradamus' famous prophecy on Louis Pasteur however, seems to defy that popularly held belief. The following is an English translation of Nostradamus' famous poem about Louis Pasteur. The poem can be found in Chapter 1, Quatrain 25 of the *Centuries*, and does in fact seem to accurately describe the life of the great scientist, even going so far as to spell his name correctly. The following is a rough English translation of the poem.

CHAPTER 1, QUATRAIN 25 (English translation)

The unseen is revealed, hidden for such a long time.

Pasteur is honored as a demi-god.

Louis Pasteur

This is when the Moon completes her great cycle (1887).

But through the slander of others, he will be dishonored.

Louis Pasteur was born in Dole, Jura, France in the year 1822, the son of a tanner. His family moved to Arbois, France when he was only two months old. Pasteur received his education at the College Communal at Arbois, but his main interests always seemed to be centered on science. He went on to study at Besancon, and entered the Ecoli School in Paris. In 1847 he graduated with a degree in Physical Science.

In 1848, Pasteur's discoveries on the refraction of light through various chemical substances won him a teaching position at the University of Strasburg as a laboratory chemist. He went on to other professorships at various universities, and in 1889 performed his research on the scientific aspects of the fermentation process as it related to the wine industry of France.

It was the 19th century, and science was rapidly emerging as the preferred method of explaining the many mysteries of the universe. In the 19th century, the fermentation process was thought to occur as the result of the "spontaneous generation" of certain molds and fungi. This somewhat religious viewpoint on the causes of fermentation developed as a result of unexplainable events often being attributed to the intervention of God. Any process that could not be completely explained or clearly understood, was usually attributed to an Act of God.

The process of "spontaneous generation" might be compared to the "spontaneous creation" theory of man's origins, as described in the book of Genesis. The scientific community had always sought more reasonable explanations for such unexplained phenomena. There was, therefore, a controversy developing between the scientific and religious communities over this issue. Some people held the religious viewpoint that such processes occurred through "spontaneous generation," while others preferred the scientific explanation. This ongoing battle between science and religion had begun many centuries earlier, as science became a more reasonable way to explain the many mysteries of God's universe.

The Church had always viewed science as a serious threat to its religious authority. The Holy Roman Church at one point had placed the great astronomer Galileo under house arrest when he had the audacity to suggest that the sun, rather than the Earth, was the center of the known universe. The Catholic Church had always taught that the Earth was the center of the universe, and that the sun revolved around it.

The idea that the Earth might revolve around the sun was not a new concept. Aristarchus of Samos expounded this controversial theory as early as 250 BC, and it was later confirmed by many famous astronomers, including Copernicus, and Johannes Kepler. The Catholic Church however, held supreme power in Europe, and would not hesitate to execute anyone disagreeing with its opinions. The Church did not admit to its error in this matter until the year 1992, when Pope John Paul II finally issued a formal apology to the scientific community for this unfortunate mistake.

When the microscope was first invented, around the year 1600, it began to reveal a previously unseen world of microscopic life forms. In the 1700's, an Englishman named Needham announced that he had "spontaneously generated" some worms (or "eels," as he called them) in a sealed jar containing putrefied animal matter.

When the French philosopher Voltaire heard of Needham's miraculous "creation" he remarked "It is very strange indeed that men should deny a Creator, and yet attribute to themselves, the power to create eels."

The religious view of the "spontaneous creation" of life was eventually challenged in many different fields of endeavor. Pasteur was able to scientifically demonstrate that the process of fermentation was actually caused by tiny living organisms too small to be viewed by the naked eye. Pasteur was both fascinated and frustrated by the hypocrisy of the great men of his time who refused to believe in what they could not see, even though they all professed to believe in an invisible God.

The invention of the microscope had revolutionized the field of science, and revealed many previously unknown secrets about the long-hidden world of the microbe. Pasteur demonstrated that the organisms responsible for the fermentation process existed in this unseen world, but were nevertheless subject to many of the same laws and rules that governed the existence of larger, more visible organisms. He also demonstrated that these tiny organisms could be killed off through the use of heat, and exposure to certain chemicals.

The wine industry of France was plagued by wine that often soured during the fermentation process. Pasteur was able to show that this problem was due to undesirable bacteria that had become involved in the fermentation process. He was able to kill off these unwanted bacteria by exposing them to high heat through a boiling process. As a result, Pasteur was able to save the wine industry of France from economic ruin.

When France's silk industry experienced an epidemic of disease in its silkworm population, Pasteur was called upon to investigate the problem. Pasteur soon discovered that only some of the silkworms carried this disease. When silkworm production was modified to separate the eggs of diseased silkworms from those of healthy silkworms, the disease was successfully controlled and eliminated.

Around the year 1880, Pasteur further extended his studies into the world of medicine. Pasteur's work was not so eagerly accepted by the leaders of the European medical community because his views went against many popularly held beliefs of that era. His work was both rejected and ridiculed by many of

Europe's top medical professionals because he dared to challenge their opinions.

Pasteur was not welcomed in many of Europe's hospitals because his views were seen to conflict with the "spontaneous generation" theory held by many of the senior members of the medical community. Pasteur however, had made many friends in the business community, and was ultimately able to fund the creation of his own medical institution known as the Pasteur Institute. Pasteur was responsible for many revolutionary discoveries in the field of medicine. He demonstrated that microbes were able to travel through the air as tiny particles, and thus transfer themselves to other hosts. He also discovered that these microbes were subject to the negative effects of heat, light and certain chemicals.

Pasteur's work in the sphere of medicine revolutionized that field. He demonstrated that the human body contained microbes of its own that could defend it against invasion by foreign organisms. Pasteur showed that the disease-causing effects of certain microbes could be weakened through outside manipulation. These weakened microbes could then be introduced into the human body, thereby strengthening the body's resistance to disease. The process of vaccination had been used for many centuries by the people of the Caucuses mountains in southern Russia to immunize their people against communicable diseases. Pasteur was able to finally explain this process, and utilize it to develop powerful vaccines to protect both humans and animals from many common diseases. Pasteur produced highly effective vaccines against cholera, rabies, and anthrax.

In one famous experiment with a herd of 50 sheep, Pasteur inoculated 25 of the sheep with an anthrax vaccine. He then introduced anthrax to the entire herd. The 25 sheep that had not received the vaccine all died; the 25 that had received the vaccine all survived. Pasteur performed similarly successful experiments with a rabies vaccine for dogs. Pasteur's experiments were always performed under precise laboratory standards of exactness and efficiency. In the year 1887, Pasteur began his work on the identification of the three most common human microbes, staphylococcus, streptococcus and pneumococcus, exactly as Nostradamus had predicted in his famous prophecy.

There can be no doubt that Pasteur's discoveries represented the most revolutionary advances ever experienced in the field of medicine. Pasteur's discovery of the microbe eventually led to the Pasteurization of milk, the sterilization of medical instruments, and widespread use of antibiotics and antiseptics. Pasteur's contributions to the surgical profession were nothing short of amazing. Surgical survival rates went from less than 50 percent to more than 95 percent through improved sanitation and the introduction of sterilization techniques. Prior to the introduction of these changes, more soldiers were killed by army surgeons, than by the enemy. In spite of all these modern miracles however, it seemed that men still failed to comprehend the nature of disease itself.

Our many successes in the field of medicine have not produced a disease-free world. In fact, quite the opposite is true. While medicine has succeeded in reducing the incidence of child mortality worldwide, the incidence of human disease has mushroomed to incredible proportions. Almost everyone you meet subscribes to a whole list of prescription drugs or medicines that they must take daily in order to survive, and genetic illnesses are now appearing as a new threat to the public health. Medical costs have also risen to unprecedented levels. The average life span for modern man has supposedly improved dramatically in the last few centuries, but a closer examination reveals that this statistic is actually due to a sharp drop in the infant mortality rate being factored into the equation.

In today's world, where God has been successfully removed from the classroom, people in the academic community are now hailing Pasteur as the man who discredited the theory of "spontaneous generation."

Discrediting religious theory was never Pasteur's intent. Pasteur was a deeply religious man, who sought only to further explain life's many mysteries through scientific investigation. He never sought to diminish religion in any way, but merely desired to explain what was not previously understood. Pasteur did not view science as being in conflict with religion. He wondered at the many mysteries of God's universe, and merely wished to promote a better understanding of them. Pasteur passed away near Sevres on the 27th of September, in the year 1895.

The prophecies of Nostradamus have also generated tremendous controversy in the world of science, as scientists still refuse to believe in what they cannot see. The long-standing battle between science and religion still rages on.

Science has always been based upon demonstrable fact, and modern scientists refuse to except anything that cannot be scientifically demonstrated or proven. The fields of religion and prophecy however, are based upon faith, not fact.

Scientists should take note of the fact however, that the following quatrain concerning Louis Pasteur was taken from a book published more than two centuries before Pasteur was born. For your convenience I've included a glossary of the Old French and Latin terms.

Chapter 1, Quatrain 25 (Old French)

Perdu trouve, cache de si long siecle.
The unseen is revealed, hidden for so long a time.

Sera Pasteur demi Dieu honore.
He will, Pasteur, as a demi-god be honored.

Ains que la lune achieve son grand siecle,

Louis Pasteur

This is when the Moon completes her great cycle (1887).

Par autres vents, sera dishonore.
Through others' slander, he will be dishonored.

OLD FRENCH AND LATIN DEFINITIONS:

Achieve - (O.F.) completes
Ains – (L.) this is, it is
Autres – (F.) others
Cache – (F.) hidden
Demi Deiu – (O.F.) half -god, demi-god
Dis-honore – (O.F.) dishonored
Grand – (F.) great
Honore – (O.F.) honored
Long – (F.) long
Lune – (F.) Moon
Par – (F.) for, through
Perdu – (F.) lost, hidden
Que – (F.) when
Sera – (O.F.) *he* will be
Si – (F.) so, such
Siecle – (O.F.) cycle, time
Son – (F.) her, its
Trouve – (F.) found, revealed
Vents – (O.F.) idle talk, slander

Louis Pasteur's work resulted in many miracles in the world of science and medicine. His name is still recognized the world over. This prophecy on Louis Pasteur is therefore one of the most famous of all the prophecies in Nostradamus' collection. It is difficult for us to deny the accuracy of this prophecy, since Nostradamus actually includes Pasteur's name, and accurately describes his work, even going so far as to include the date of the accomplishment (at the end of the Roussat Moon cycle, 1535-1887).

Prophecies such as this one have caused people all over the world to sit up and take notice of the works of this amazing Hebrew prophet. His prophecies are also interesting because they were written a mere 450 years ago, and are therefore not as difficult to translate as Bible prophecy. They also seem to agree with Bible prophecy, and in many cases, help to further explain it.

Nostradamus' prophecies were written in a relatively modern language that is not as difficult to translate as the ancient biblical languages, whose limited vocabularies allow for a wide variety of interpretations. They were also not subject to multiple inter-language transcriptions down through the centuries as is the case with Bible prophecies.

For these reasons and many others, the prophecies of Nostradamus, the most recent of the 500-year Hebrew prophets, have captured the attention of the entire world, and are still a focus of international attention every time a major world event occurs.

CHAPTER EIGHTEEN:
THE FLOOD

And God saw that the wickedness of man was great in the earth, and that every imagination of the thoughts of his heart was only evil continually. And it repented the Lord that he had made man on the earth, and it grieved him at his heart. And the Lord said, I will destroy man, whom I have created, from the face of the earth....

(Gen. 6:5)

ONE OF THE BEST KNOWN of all the Old Testament stories is the story of Noah and the Flood. This narration is found in the book of Genesis Chapters 6 through 8. Most of us are thoroughly familiar with this story describing how Noah and his family and a great host of animals were saved from the devastation caused by this great catastrophe. Many of these early Bible stories have often been viewed as being figurative in nature, serving only to prove a point. Not many biblical scholars view them as literal happenings.

The writings in the Old Testament represent the best efforts of early Bible writers to translate these ancient stories into modern language. These stories therefore, leave much to be desired when it comes to exact literal interpretation. Assuming that there is some underlying truth to the story of the Flood, it still remains for us to determine through scientific means, whether or not this event could have actually taken place.

The story of the biblical Exodus was also viewed in the same perspective as the Flood. Recently however, it was established through scientific means that the timing of the biblical Exodus exactly coincided with the explosion and sinking of the great volcanic island of Stronghyli in the eastern Aegean Sea. It is a scientific reality that the tidal wave created by the collapse of this island would have most certainly caused the Red Sea Canal to be completely emptied in advance of the arrival of the great tidal wave upon the shores of Egypt. The only question that remains for us to answer is how Moses was able to so accurately time the event.

The biblical Flood on the other hand, is a catastrophe on a much greater scale than that of the parting of the Red Sea, and would therefore require a much greater natural cataclysm to explain it. We do know for instance, that there is not enough water available on Earth to cover the entire surface of the planet. So how could the biblical Flood have possibly occurred, and is there any scientific evidence for such a flood?

One famous archeological discovery that does seem to lend some credence to the Flood theory, was a discovery made in 1929 by a British archeologist named Sir Charles Leonard Woolley. Woolley was excavating in the ancient ruins of the Mesopotamian city of Ur, when he made one of the most amazing archeological finds of all time. He and his workmen were digging in a 5-foot square vertical shaft in a cemetery near Ur, when they suddenly came upon a layer of water-laid mud that contained absolutely no archeological ruins.

Woolley's diggers assumed they had finally struck virgin soil, and ceased to dig any deeper into the shaft. Woolley however, was bothered by the fact that the archeological rubble had ended so soon. He therefore ordered his diggers to keep on digging.

The Flood

The diggers begrudgingly complied with Woolley's request, and after removing another 8 to 10 feet of this mud, they suddenly came upon more archeological ruins.

Puzzled by the presence of this unusually thick layer of clay, Woolley revisited the excavation the next day with his wife, who walked away remarking that this must of course be the great Flood. The mud after all, was located at the 2500 BC level of the excavation.

Woolley was absolutely floored by the prospect that this layer of mud might actually be physical evidence of the great Deluge, but what other explanation was there? After all, he had just excavated almost 4500 years of continuous human existence at this location, and there was no other plausible theory for this unexplained layer of water-laid clay. Could it be that the biblical Flood actually did take place?

As with many other religious questions, it would probably be best for us to look to the Bible for the answer. As we read the book of Genesis, we may notice a few unusual things about its early chronicles. For instance, when Cain, the son of Adam, slew his brother Abel, and was banished to the land of Nod, Cain begged the Lord not to send him to Nod because he feared that everyone who found him there would slay him. The question is, exactly who was "everyone"? The Bible tells us of only three people living on the Earth at this time. So who were these people who were going to slay him? Is it possible that there were other people living on Earth at the same time as Adam, Eve, and Cain?

This same question comes up once again in Genesis Chapter 12. In this narration, describing the dissemination of the sons of Noah and their families throughout the area around Iraq, Abram decides to leave the land of his kin and sojourn southward into the land of Egypt in order to avoid a famine that has struck his land. He sojourns south into the land of the Egyptians. The question then logically arises as to where these Egyp-

183

tians came from, since they are not counted as the descendants of any of God's people.

These Bible references, and others as well, raise the possibility that perhaps the biblical Flood may have been confined to only one specific area of our planet. God, after all, was only trying to cleanse the Land of the Patriarchs, not the entire Earth. This theory would certainly make a lot more sense from a scientific point of view. If this theory is correct, then the question arises as to what sort of an occurrence could have accounted for a catastrophe on the scale of the biblical Flood?

There is one scenario that might explain this sort of a disaster. We know for instance, that the motion of the ocean tides is caused by the gravitational forces of the Moon drawing up the oceans onto one side of the Earth, resulting in the daily rising and falling of the tides. This gravitational force is extremely powerful, in fact, in some areas of the Bay of Fundy, the ocean is known to rise and fall as much as 46 feet in a single day.

It might be possible to draw the world's oceans up onto one side of the Earth and produce conditions similar to those of the great Flood, if a strong enough gravitational force could be generated on one side of the Earth. The conditions might have been created if a large heavenly object had passed by very close to our planet. These conditions would have been particularly pronounced if the object had passed by on the same side of the Earth as the moon was positioned in at the moment. These circumstances would most certainly have produced a high tide on one side of the Earth capable of completely submerging a large portion of the globe. The meteorological disturbances caused by such a planetary near-miss would have lasted for many weeks.

If, in 2500 BC, a large, unidentified object from space did pass by very close to Earth, then the Flood may have occurred exactly as the Bible describes. This massive astronomical object might even have been an undiscovered planet within our own solar system. This planet could possess a widely eccentric orbit that brings it around the sun only once every few hundred years or so. As such, it might not have been noticed by early astronomers here on Earth. This theory would fulfill all the scientific aspects of the physical conditions necessary to produce the massive flood described in the Bible.

It is much too soon for science to subscribe to such a theory, but you never know, someday this object may decide to make itself known by coming around once again and taking another shot at us!

CHAPTER NINETEEN: EVOLUTION

"If the misery of the poor be caused not by nature, but by our own institutions, then great is our sin."

Charles Darwin

THE THEORY OF EVOLUTION has been a source of great controversy in the religious world for many decades. The person credited with developing this controversial theory is a British naturalist by the name of Charles Darwin. Darwin was a university-educated scientist, who at the age of 22, decided to circumnavigate the globe in order to study nature on the world's major continents.

In December of 1831, Darwin set sail aboard the HMS Beagle on a five-year voyage around the world. Darwin was originally planning to make religion his chosen field of endeavor, but he had some questions about the true nature of Creation, and wanted to settle them with this trip.

When Darwin returned from his famous excursion in October of 1836, he immediately canceled his plans to go into the church, and decided instead to further explore the world of science. During his long cruise, Darwin had visited a group of tiny islands off the western coast of South America called the Galapogos Islands. On the Galapogos, he documented the existence of many strange and unusual animals that differed greatly from similar species on the nearby South American mainland. Darwin was convinced that these remarkable differences resulted from a process he called Natural Selection, that is, survival of the animals most fit for their particular environment. Darwin theorized that animals less fit for their environment were eliminated through this natural process, and thereby prevented from passing on their inferior genetic traits to the gene pool of their particular species.

Actually, Darwin was not the first person to recognize this natural process of Survival of the Fittest. This phrase actually came from another British naturalist named Alfred Russel Wallace, who wrote the phrase down in a letter he sent to Darwin long before Darwin published his famous theory. Actually both these men were outdone by America's great scientific genius, Benjamin Franklin, who also described the process many years before either Wallace or Darwin.

This line of thinking was certainly not new. The science of genetics had been recognized and utilized by man for thousands of years. The ancient Chinese for example, had noticed that some of their carp exhibited a faint color when held up to the sunlight. They found that when they bred these carp together, they were able to produce fish of even brighter color. They then destroyed any fish of poor color, and bred only the most brightly colored fish together. After hundreds of such breeding cycles, they were finally able to produce the breed of carp we know today as the goldfish. The Chinese utilized this same selective breeding process with many other animal species as well, in order to please the refined tastes of their emperors.

In more recent times, this process of genetic selection, whether natural or artificial, has also been observed and utilized by man. Modern farmers for instance, often breed their animals for specific traits. Milk farmers will selectively breed their cows to produce more milk, while beef ranchers breed their cattle to gain more weight. Dog breeders for many centuries have selectively bred their animals for specific tasks like hunting or herding sheep.

The science of animal husbandry has long been a useful tool for improving certain animal species. It was always understood that genetically inferior animals needed to be destroyed as a normal part of this process. Natural Selection has always been recognized as a powerful tool used by nature to refine its various animal species.

Darwin theorized that it must have been this same process that was responsible for the genetic improvement of all species, including man. Darwin thought that modern man must have evolved from the lesser apes into the present-day species known as homo sapiens. Darwin's theory infuriated many people in the church world, particularly those who believed in the literal translation of the book of Genesis that described God's "spontaneous creation" of man. Even those who thought the creation story to be a figurative tale, had great difficulty accepting the idea that they were originally descended from apes.

Darwin's controversial theory found favor however, with many of the liberal-minded members of the academic community, who began to teach the controversial theory in America's public schools. Eventually a legal battle developed between Evolutionists and Creationists. The battle culminated in the famous "Monkey Trial" of 1925, to prevent the teaching of evolutionary theory in public schools.

Tennessee schoolteacher John Scopes had defied a state law preventing the teaching of evolution in Tennessee public schools. The Scopes trial made newspaper headlines all around the word, and resulted in a temporary victory for the Creationists, and a $100 fine for Scopes. Scopes' conviction however was eventually overturned in the Tennessee Supreme Court.

Evolution

Many liberal educators continued to teach the "science" of evolution in America's schools, as scientists everywhere anxiously sought evidence to support Darwin's controversial theory. Archeological digs were initiated all over the world to locate the many "missing links" between man and ape. Every major science museum had a set of human-like figures on display, representing the many gradual changes taking place in the evolutionary development of man from lesser, ape like, creatures.

There were many important indicators that scientists used in determining where each new archeological find was to be placed on the evolutionary scale. One indicator, canine tooth development, was considered to be extremely important. Chimpanzees, man's closest relative in the ape family, had long, well developed canine teeth, as opposed to humans, who exhibited only slight canine development. It was therefore thought that the canine teeth of modern Chimpanzees were a holdover from a time in the distant past when Chimpanzees were meat eaters instead of herbivores.

In the early 1960's however, the scientific community was shocked to its core, when a group of wild Chimpanzees were observed eating a small monkey. Further investigations revealed that these chimps had actually hunted down and killed the monkey in order to eat it. Evolutionists were greatly embarrassed to find out that the Chimpanzee's long canine teeth were not a vestige from the past after all, but instead served a useful purpose in their present-day diet.

Developments from another archeological find in Africa made matters even worse for the Evolutionists. In 1976, an archeological find in Tanzania revealed the existence of early hominids that walked upright on two feet, just like modern human beings. This rare archeological find was dated at over three and one half million years old, just about as far back as you could get on the human evolutionary scale. But when scientists examined these hominids for canine tooth development, they

189

found absolutely none at all! Their teeth were as normal as the modern human teeth of today, exhibiting not even the slightest hint of canine tooth development. This discovery put a big crimp in Darwin's theory. Up until now, all "missing link" finds had possessed large canine teeth, and all were dated at less than two million years old!

Another important indicator of human evolutionary development was the construction of the foot. Apes had feet shaped more like hands, since they were used for grasping tree limbs while climbing. It was therefore thought that the anatomy of the feet of these "missing links" would also exhibit a gradual evolution from the hand-like foot of the ape, into the modern human foot structure of today, designed for walking upright. But the feet of these almost 4-million year old creatures from Tanzania were exactly like those of modern human beings, with not even the slightest hint of the thumb-like big toe so characteristic of all monkeys and apes. Also, the design of their hip joints indicated that these early hominids walked fully upright, just like modern human beings!

Suddenly, all the scientific evidence was leaning toward Creation Theory. The Creationists had long held that evolutionary influences were valid as long as they concerned only the refinement of a species, but were invalid when they concerned the creation of species. It was now clear that tooth shape was determined by diet, and foot shape was determined by mode of travel. The Creationists had always taught that creation of species was through the will of God, and that evolution merely served to further adapt a species to its particular environment.

The Theory of Evolution was causing some other problems as well. It seems that modern advances in the field of medicine were beginning to have an adverse effect upon the process of human evolution. The genetically weaker members of human society, formerly eliminated through the process of natural selection, were now surviving due to many technological advances in the field of medicine. Evolutionists were concerned about the future genetic integrity of the human race itself; after all, the genetically weak were now surviving to pass on their inferior genes to the general population. It was now a fact that the genetically inferior members of the lower classes were breeding at a much faster rate then the genetically superior upper class. The process of evolution had now been essentially reversed! Evolutionists therefore felt it their duty to take immediate action to eliminate this threat to the genetic evolution of the human species, and so a new science, called Eugenics, was born.

Supporters of this new Eugenics movement lobbied the government to seek out the genetically inferior members of human society and take immediate action to ensure that their genes were not passed on to the general population. In America's Midwest, it had long been common practice to sterilize the mentally ill, or "mental defectives" as they were called, in order to prevent them from having children. This practice enjoyed the widespread support of the government; in fact, the State of Indiana in 1907 enacted a law that legalized sterilization of the mentally ill.

Not much happened with the Eugenics movement until around the year 1920, when the economic and social successes of that era gave rise to a rebirth of liberal idealism, and generated new support for social change in order to improve human society. Drunken with the economic successes and technological achievements of the Industrial Revolution, wealthy liberals now set their sights upon improving the genetic integrity of the human race itself, through their new science of Eugenics.

The Eugenics movement enjoyed the enthusiastic support of the members of white, upper class society, who felt that their superior intellect endowed them with the responsibility to make such decisions. Eugenics supporters overwhelmingly favored social engineering for the lower classes in order to protect the racial integrity of America.

The Eugenics concept, now renamed "Social Darwinism," was heartily embraced by America's intellectual elite as holding great promise for the future of all mankind. The United States government would now be able to address the many issues that threatened the genetic purity of the nation. Issues such as immigration would need to be addressed immediately. Immigrants to America's shores would now have to be screened for mental IQ, and other physical factors, before being allowed to enter the country. America's genetic purity could not be compromised by the genetically inferior castoffs of other nations. Strict mental and physical standards would need to be established for all future immigrants.

The many accomplishments of America's Eugenics movement did not go entirely unnoticed. The members of Hitler's Nazi party in Germany were very impressed with the many successes of America's Eugenics movement. The Nazis had long been seeking a workable program to help them maintain the genetic purity of their superior Aryan race.

Darwin's theories had been popularly supported by many Germans in the First World War through the writings of the German biologist Ernst Haeckel, who justified the right of the superior Caucasian race of Germany to lead the way in the development and improvement of Aryan society. Social Darwinism now promised to be the key to unlocking the full potential of the German people.

The original seed planted by Darwin had now produced its ultimate fruit. As the Serpent had so long ago whispered to Eve, "in the day that you eat of the Fruit of Knowledge, your eyes shall be opened, and you shall be as gods." Now, with this godlike power endowed through knowledge, Hitler and his Nazis could finally realize their dream of creating the perfect human society. The Enlightenment had now given birth to another great world-conquering beast.

Evolution

As had occurred so many times in the past, the dream of the liberal elite for a society unfettered by the laws of God, was once again loosed upon mankind. Millions of followers of the new Messiah, Adolph Hitler, failed to realize that such men are driven only by their own insatiable lust for power. The greatest of all the beasts would not now be satisfied until he had claimed more human lives than any beast before him.

Darwin's name could now be added to the long list of social reformers, like Karl Marx, Vladimir Lenin, and the great Voltaire (Jean Francois Arouet), whose liberal ideas on how the world should be run eventually brought them into conflict with God. Their idealistic dreams for a better human society had given birth to men like Napoleon and Hitler, who sought to bring their sinful ideas to fruition. Like small children playing with matches, foolish men believed that knowledge and its resultant technologies would ultimately allow mankind to create the perfect human society.

The ancient prophets taught that man's only path to Eden was always through belief in, and subservience to, the supreme laws of God. Man's quest for Eden, launched by Eve's original sin, had brought only death and destruction to the world. For over 6000 years, it seems that man had been attempting to achieve what God had already freely provided.

CHAPTER TWENTY:
THE FIRST THANKSGIVING

"Thus out of smalle beginnings, greater things have been pro-duced by His hand that made all things of nothing, and gives being to all things that are; and as one small candle may light a thousand, so ye light here kindled hath shone to many, yea, in some sorte, to our whole nation."

William Bradford

THE ABOVE VERSE WAS WRITTEN many years ago by William Bradford, Governor of the Pilgrims of the original Plymouth Colony in Massachusetts. Don't be fooled by the Old English script, the letter "y" was actually pronounced like our modern English "th" sound. That's right, "Ye Olde Antique Shoppe" is actually just "The Old Antique Shop."

Not many modern Americans are aware of the true reasons why we celebrate our Thanksgiving holiday. I therefore thought you might enjoy hearing this true story about the interesting events that led up to the establishment of this uniquely Ameri-can holiday. We owe a great debt indeed to the small group of English Pilgrims whose struggles in the New World paved the way for the Christian colonization of America.

The First Thanksgiving

In the first week of September, in the year 1620, a small band of Christian Protestants set sail from Plymouth Harbor in England on a long journey across the Atlantic Ocean. Their intention was to establish a new Christian society in North America. Their trip was fraught with many perils and storms, but the good ship Mayflower held together well, and in a few weeks they began to sight small groups of water birds feeding on schools of fish. This was a good sign, for it meant that land could not be far away.

On the morning of November 9th (Old Calendar) at around daybreak, the crew of the Mayflower caught their first sight of land in over two months. The narrow strip of land they sighted was known on the sailing charts as Cape Cod. The Mayflower immediately altered its course to seek out a river known to lie about 10 miles south of the Cape, but the winds were against them and Captain Jones decided it was best to head for the safety of the Cape Cod (Provincetown) bay. This bay provided shelter for great flocks of water birds on their annual flight south for the winter. Never before had these Christian Pilgrims seen so many birds in one place at one time. It was a truly magnificent sight to behold. Every day aboard ship, they were treated to the sight of great whales frolicking about in the bay. The Pilgrims had no means aboard their ship to harvest these whales, which could have been a good source of food, and also valuable whale oil.

The Pilgrims were sorely in need of replenishing their dwindling supplies of firewood and fresh water. The Cape Cod bay was very shallow, especially near shore, and it was necessary for them to wade great distances through the cold water in order to reach the shore. This proved to be a serious problem in cold weather, for it caused many in the group to become ill. It was decided that sixteen armed men would be dispatched to shore to gather the necessary supplies. The men embarked upon the cape side of the bay.

Trail of Prophecy

The cape turned out to be a rather narrow strip of land, bordered on one side by the bay, and on the other by the open sea. The shoreline of the New World was heavily wooded with mature oaks, pine, juniper, sassafras, and other sweet woods. The edge of it consisted of endless sand dunes. The interior of the land was thickly forested by trees with foliage so dense that not a trickle of light penetrated through. Without light, there were no trees or underbrush growing beneath this forest canopy. It was open and unrestricted underneath so that a carriage could be freely driven about under it. Its soil was rich and black. Never before had the Pilgrims seen a land so rich as this place.

A few members of the group decided to venture ashore to repair the ship's small sailing skiff, which was used to ferry them to and from the shore. The repairs took longer than intended however, and it was decided that a group of armed men should be sent out to see if they could contact any of the native inhabitants of the area.

A group of sixteen armed men, under the direction of Captain Miles Standish, was sent out along the shore to see what they might find. After traveling only about a mile, they spied a group of 5 or 6 people with a dog, walking toward them along the beach. The Pilgrims shouted to these natives, but the savages quickly ran into the woods and called their dog in after them. Standish and his men rushed into the woods after them. They knew they had no hope of catching up to the natives, but thought perhaps they might follow them to a nearby village. After tracking them for nearly ten miles, they saw where the savages had climbed to the top of a hill to see if they were being followed. With heavy armor weighting them down, the exhausted Pilgrims decided to set up camp for the night.

In the morning, the party once again picked up the native trail and followed it farther into the woods. After a short time it led them into a thick patch of briars and tangles, and it soon became evident that they had been deliberately led on this wild goose chase. The party continued on however, and eventually came to a small valley filled with deer trails. These trails led to springs of fresh water where the thirsty Pilgrims had their first drink of icy cold New England water. They drank it with great delight, for they were tired and thirsty, and sorely in need of this refreshment.

The men then decided to head farther south in order to pick up the shoreline again. When they finally reached the waters edge, they lit a signal fire to let the Mayflower know of their new location. They then decided to explore another nearby valley, where they soon stumbled upon a fresh water pond about a quarter mile in length. Near this pond they located an open field of approximately 20 acres, that had been previously cultivated by natives. Not far beyond this field they encountered a strange mound of earth that was covered with straw matting, and topped with a large wooden mortar. At the end of this earthen heap, they found a small native cooking pot sitting in a hole that was dug into the ground. Some members of the party decided to dig into the mound, and soon located a partially rotted set of bow and arrows. It then occurred to them that this might be a native grave, so they restored it to its original condition, and continued on their way.

After traveling a bit farther, they came upon a small field that was filled with corn stubble from this current year. Passing by another opening, they came to a place where a native house had once stood, with four or five wooden planks laid together on the ground and an old iron kettle from some ship's galley. They then spied another heap of earth, but this one was much smaller than the first, and not covered with any matting or decoration. They dug into this mound as well, and uncovered a native basket filled with corn from a previous year's harvest. Digging even deeper, they came upon another basket containing fresher corn from the current year. It was a large basket of Indian corn, some yellow, some red, and some mixed with blue, a very welcome sight to the weary travelers. They decided to take the corn with them, and also the kettle, and later return the kettle to the natives, and settle with them for the corn.

Not far from this place, the party discovered a large wooden palisade. The palisade was located near where a river was supposed to be. The men quickly located the river and then followed it back to the open sea. Along the way they came upon a native canoe hidden on the riverbank, and also spied another canoe on the opposite bank. They once again set up camp for the night, and posted a guard. It was raining, and the group tried to sleep as best they could in the cold and wet conditions.

The following morning they were able to locate another trail that led to the east, and soon came upon a young sapling bent down over the trail, with a pile of acorns on the ground underneath. One of the men said it was a native trap to catch deer. When William Bradford, following up from the rear, tried to walk around the trap, it sprang up and caught him by the leg. After freeing Mr. Bradford, the party marveled at the fine quality of the rope, and took it along with them as they continued on their way.

When they finally reached open sea again, they had some distance to travel before they could sight their ship. They then fired off a few signal shots, and a lifeboat was dispatched to pick them up. The ship's skiff was now fully repaired, but it was decided that another skiff needed to be built, and so the ship's carpenter was set to that task. The weather was now growing much colder, and trips to and from shore were becoming a real problem. These trips could only be undertaken at high tide, and the group had to wade great distances to shore in thigh-deep water, which caused many of them to catch coughs and colds that later resulted in many deaths.

When the new skiff was finally completed, a second expedition was undertaken to further explore the native settlement where the corn was found. This time, thirty-four men, including ten from the ship's crew, set sail in the lifeboat and skiff to relocate the river where the native village was found. The weather on this trip was very bad. It was snowing and the wind was blowing fiercely. The skiff soon began taking on water, and had to be sent ashore. The men in the lifeboat continued on, but the wind blew all that day and evening, and more men fell ill from this trip. The bay (Truro bay) was finally located, but it was found to be much too shallow to accommodate such a large ship as the Mayflower.

The group then decided to further explore the site where the original native village was found, to see if more food could be located. They discovered the native canoe still resting on the bank of the creek, and also sighted a large flock of geese. The men shot six of the geese and fetched them with the canoe. They also made use of the canoe to cross the river. It was a large canoe, and could easily carry 7 or 8 men at a time. The party soon relocated the native village and proceeded to dig up more earthen mounds that were found to contain large stores of corn, wheat, and beans.

The Pilgrims thanked God for guiding them to these large stocks of food and grain, for without them they would not have survived their first winter in the New World. The ground was now covered by snow 6 inches deep, and also frosted to a depth of 6 inches, so it was nearly impossible to dig into it. Many of the men were now feeling ill, and decided to return to the ship. About half of the group stayed on to further explore the area.

They soon located another earthen mound, but this one was much larger and much more elaborately decorated than any previously discovered. It was found to contain the body of a person with blond hair, and also the body of a small child. The party was surprised at this unusual find, because it was known that the native peoples all had black hair. Was this a Christian who had lived among them?

A further search of the area yielded a small group of native houses that had been recently dwelt in. The houses were constructed of green saplings bent into an arch, with both ends stuck into the ground. The houses were round, and covered on the outside with matting. The matting consisted of two layers, with the inner layer being of a much finer weave and much newer than that on the exterior. The door to each house consisted of a flap about 30-inches square, and there was a hole in the roof that served as a chimney. One could stand fully upright inside the structure, and in the middle of the floor there was a pit for a fire. There were double layers of matting laid round in a circle inside for sleeping, and outside there were piles of reeds for weaving more matting. There were also numerous baskets, bowls, trays and dishes, all filled with various sundry items including clams, crab shells, acorns, and pieces of smoked herring and venison.

It was now December, and winter was rapidly closing in. A decision had to be made soon about where the group would settle. Some wanted to settle at the present location because of its cleared land and fresh water, but others wanted to explore an even larger bay, called Agawam, that was said to lie north along the coast. It was thought that Agawam might be unsuitable however, because it was too large and too far away, and so a third expedition was launched to see if a more suitable nearby location could be found.

An expeditionary group was sent out in the lifeboat and skiff to further explore up the coast. The weather was extremely cold and damp, and the men's clothing was soon frozen hard from the cold sea spray. After traveling approximately 40 miles along the coast, they spotted a group of savages on the beach gathered around a large black object. It was late in the day, and so the party decided to set up camp for the night. They could see the native campfires burning about 5 miles in the distance.

The First Thanksgiving

In the morning, they set out for the location where they had spotted the natives, and soon came upon a grampus, or small whale, that had been washed up on the beach. When they arrived at the location where they'd seen the natives, they saw where the savages had been stripping the body of another grampus. There were numerous bare footprints all around the carcass. The party then decided to travel inland, and soon came upon a large wooden palisade, which contained many native graves. Some of these graves even had houses built over them. The hour was growing late though, and so they decided to set up camp for the night.

At around midnight, a loud, whooping cry was heard coming from the forest. The men shot off their muskets and the sound stopped. The next morning around daybreak, they again heard the whooping cry, and this time arrows were seen to fly at them from the woods. They could plainly see one of the savages standing about a half a musket's shot away. He shot three arrows at them, but the men ducked his arrows and returned a round of musket fire, wounding him in the arm. The savage and his men then ran off into the woods. The Pilgrims picked up some of the arrows and continued on their way.

The party set out to sea once more to explore further up the coast. After traveling another 40 miles, they ran into some bad weather, but continued on, and were at the last minute able to find shelter in another large bay (Plymouth bay). They spent the night on an island in the middle of the bay, and in the morning were able to determine by sounding that this bay was indeed suitable for large ships. Further explorations of the area revealed many cleared fields and numerous sources of fresh water; so the party quickly sailed back to the Mayflower with good news of their find. In the third week of December, the Mayflower finally set anchor in Plymouth Harbor.

Stormy weather continued to plague the weary travelers, but in a few days they were finally able to go ashore to begin cutting the lumber they needed to build their new settlement. They chose to build the settlement upon a high hill that would be defensible from savages, and also offer good views of the harbor and the open sea. The Pilgrims often saw native campfires burning in the distance, but the savages never approached the settlement.

The area was a perfect spot for a farming settlement. There were many cleared fields and numerous fresh water streams filled with fish. There were also cherry trees, and plums, and strawberries, as well as great stores of leeks and onions. By January, construction of the settlement was well underway.

A few days later, two men from the settlement were walking their dogs in the forest when one of the animals chased after a deer. The men soon became lost in the woods and were forced to spend a night in the forest. They found it necessary to climb a tree in order to escape hungry mountain lions roaring nearby. Predators in the New World were extremely plentiful. An area only 5 miles square would often contain as many as 100 wolves, 40 mountain lions, and 20 black bears.

The next morning, the men ascended a high hill and successfully relocated the bay. They eventually arrived back at the settlement cold and frightened, but none the worse for their adventure.

The following week, Master Goodman was walking his spaniel in the woods when the dog was accosted by two large wolves. The dog ran between its master's legs for protection, and Master Goodman held the wolves off with a large stick. The wolves soon tired of the standoff and retreated back into the forest.

A day or two later, another member of the group was hunting in the forest, when he spotted a group of savages headed in the general direction of the settlement. He waited until they had all passed, then ran back to sound an alarm. The workmen, who were in the forest, ran back to the settlement and armed themselves. The natives however, never showed. When the workmen returned to the forest, they found that many of their tools were missing. The Pilgrims resolved never again to assume that they were not being closely watched by the natives.

Many native houses were eventually located in the area, but none had been recently dwelt in, and the Pilgrims were unable to successfully establish contact with any of the local natives. Winter was now upon them, and food was growing scarce. Their first winter in the New World was a severe trial for the Pilgrims. Food supplies were dwindling, and many were dying from colds and pneumonia. More than once their hastily built cabins were set afire by errant sparks from the fireplaces, and an attack by the savages was a constant fear. In their present condition, a single attack could have easily overwhelmed them, but that attack never came. Of the 102 original members of the group, only about 60 survived their first winter in the New World.

By March 1st, the snow was finally melting, and the first signs of spring were beginning to appear. Then, about two weeks later, on a Friday, a tall, unclothed savage boldly walked up to the gate of the settlement and greeted the Pilgrims in English. He said his name was Samoset, and he spoke a type of broken English he'd learned from the many English vessels that visited the area. He said he was not of these parts, but was of the Moratiggon, a tribe to the far north, and one of the sagamores thereof. He was a tall man with long black hair falling behind, and short hair in front. He had no hair on his face at all, and was stark naked except for a leather skirt bound about the waist with a fringe at the bottom. He said that he'd been in these parts about eight months. The Pilgrims greeted him as a friend and invited him into the settlement, but all the men kept their firearms at the ready.

Samoset proved to be an excellent source of information concerning the area. He told them that the place they now occupied was called Patuxet, and that it was previously inhabited by a native tribe, totally wiped out by a plague that had come through 4 years earlier. The plague killed all the Patuxets, and about 90 percent of the members of other tribes in the area. He also told them that their land was therefore free for the taking. The Massasoits, their neighbors to the west, who once numbered nearly a thousand, now numbered only about sixty, and the Nansets to the southwest, were about a hundred strong. The Nansets were the tribe that attacked the Pilgrims in their first unfriendly encounter.

All the tribes in the area were angry with the British because of the English Captain Howe, who had deceived them and sold about twenty of their number into slavery to the Spanish. The natives had recently killed three of the Englishmen in retaliation for this despicable act. The Pilgrims sincerely apologized for Howe's behavior, and assured Samoset that not all Englishmen behaved in this manner. Samoset was then entertained with food and drink, both of which he heartily enjoyed. The following day he returned to the Massasoits but promised to return soon with other natives to trade animal skins.

The very next day Samoset returned with five other savages who were also very tall, and clothed in deerskin trousers. Their faces were painted black, and they wore feathers in their hair. Their chief carried a wildcat's skin over one arm and a fox-tail in his hair. They left their weapons about a quarter mile outside of the settlement as had been previously requested of Samoset.

The Pilgrims fed and entertained their native guests, and the natives likewise danced around the campfire in their strange antic manner. They had brought only five animal skins with them however, so the Pilgrims sent them away to fetch more skins, and also requested that they return the tools they'd stolen from the woods. The natives returned the tools the very same day, and promised to return soon with more animal skins.

In a few days Samoset returned once again, this time accompanied by another savage named Tisquantum (Squanto), who was the only surviving member of the Patuxets. Tisquantum had been taken captive by Captain Hunt, and sailed to England as Hunt's servant. Tisquantum thus escaped the plague that killed his tribe. The two men told the Pilgrims that the great Chief Massasoit was encamped nearby, and wished to meet with them. The Pilgrims agreed to meet with the chief, and an agreement with Massasoit was soon concluded over a hearty meal and some strong drink, which caused the chief to sweat profusely all during the meeting.

When spring finally arrived, the planting of some 20 acres of corn, wheat, and peas was undertaken in the native manner using alewives (American Shad), which clogged the streams and rivers each spring, as fertilizer.

Through a policy of Christian fairness, honesty and respect, the Pilgrims were finally able to establish a peace with, and among, the various native tribes in the area. Many of these tribes had formerly warred with one another, but now a state of peace prevailed throughout the land, and it was possible for any man, Christian or native, to travel the forest trails in complete safety. The natives freely visited the homes of the Pilgrims, and great knowledge was passed on concerning native customs, foods, and medicines. The colonists had applied Christian principles to create peaceful and positive change in a people who had for thousands of years been suffering in ignorance. Never before had these native tribes lived in such a state of peace and mutual respect with one another.

In the late summer, a young boy from the settlement became lost in the woods and could not be located by search parties. Word eventually came to the settlement that the boy had been picked up by the Monomets and taken to the Nanset village to the south. The Nansets were the tribe who had attacked the Pilgrims in their first unfriendly encounter, and it was therefore feared that they might do harm to the boy. A group of ten armed men from the settlement was therefore sent out by boat for Nanset, accompanied by native interpreters, in order to safely retrieve the boy.

The trip was interrupted by a severe thunderstorm, and the group sought refuge in Comoaquid (Barnstable) harbor. The Comoaquid chief, Iyanough, agreed to accompany the Pilgrims on their trip to the Nanset village. They arrived at Nanset late in the day, and the native interpreters traveled to a small nearby village to inquire about the boy. The Nanset chief Aspinet returned with about a hundred of his tribe and returned the boy safely to the Pilgrims. It was said that the boy had wandered for 5 days before being picked up by the Monomets.

Aspinet told the Pilgrims that a sachem named Corbitant was plotting with the Narragansetts against the Pilgrim settlement, and that he had taken the great Chief Massasoit prisoner, and also Tisquantum. The Pilgrims were greatly disturbed by this news and decided to immediately undertake an expedition to Namasket to rescue both Tisquantum and Massasoit.

The Pilgrims sent a party of ten armed men to Namasket. The men arrived late in the evening and found Tisquantum alive and well. They shot and wounded a few members of the Namasket tribe, and Corbitant and his followers fled into the woods. Tisquantum was able to calm the rest of the tribe, and the Pilgrims left a warning that Chief Massasoit must be returned unharmed. They also warned the villagers that any natives found to be siding with Corbitant would be hunted down and killed.

The balance of the summer went very well for the Pilgrims, and their first harvest in the New World was a bountiful one. In October the governor of the settlement decided that a great feast of Thanksgiving should be held to thank the Lord for all their good fortune.

Hunters from the settlement were sent out to shoot wild turkey and waterfowl for the great feast. The women gathered large quantities of grapes and plums, and the natives were also invited to attend. The natives brought with them baskets of clams, oysters, native pumpkins and popping corn. Even the great chief Massasoit attended with about 100 of his men, and provided five deer for venison. The great feast of Thanksgiving continued for three days, and was heartily enjoyed by both natives and Christians alike. And so Thanksgiving was thus established as the first Christian holiday in America.

It was a truly happy time for the Christian Pilgrims in this new well of plenty. God had paved the way for them with a plague upon the savages, and had also led them to settle upon the only 20-mile section of coastline in the entire New World unclaimed by any native tribe. Truly the Lord does work in mysterious ways.

CHAPTER TWENTY-ONE: EASTER

"And (Thomas) Morton became lord of misrule, and maintained a school of Atheism. They also set up a May-Pole, drinking and dancing about it many days together, inviting the Indian women for their consorts, dancing and frisking together like so many fairies (or furies rather), and worse practices."

William Bradford - Governor, Plymouth Colony

THE EASTER HOLIDAY is still celebrated every spring by Christians honoring the resurrection of Jesus Christ. Easter is therefore one of the most meaningful of all Christian holidays. Easter is commonly celebrated with Easter bunnies, Easter chicks, Easter eggs, and Easter sunrise services. Easter is also preceded by a period of fasting known as Lent.

The tremendous symbolism of the Easter holiday still remains a mystery for most Christians, but it is important for Christians to explore that symbolism if Easter's true origins are to be revealed. The Easter celebration was passed down to us from the traditions of the Holy Roman Church of Europe. The Catholic Church rose out of the Holy Roman Empire, which was originally formed through a union of the Roman Empire and the Christian Church. This union was first formed around the year 313 AD, when the Roman emperor Constantine struck a deal with the Christians to have Romans accept the Christian faith. Constantine claimed to have received a vision of a cross in the sky, bearing the words "In hoc signo vinces" (by this sign thou shalt conquer).

At the time this deal was originally struck, the Roman Empire was in a state of serious decline, and Constantine had to act quickly to strengthen his hold on the European continent. The Christians had made huge inroads into northern Europe, and also controlled a major portion of the Roman Empire itself. Constantine therefore decided it was in his best interests to form a union with the Christian church in order to prevent his empire from crumbling.

As part of this agreement, the Church agreed to change its Sabbath from the seventh day of the week to the first day of the week. The Pope of Rome agreed to this compromise in order to accommodate the wishes of the pagan Roman sun-worshippers. The Christian Sabbath therefore, was officially changed to Sunday, the Roman day of the sun. In exchange for this compromise, the Church would be protected by the Roman government, and would also have a voice in how that government was run.

It was advantageous for Constantine to strike this deal with the Christians, because it extended the boundaries of his empire into the farthest reaches of barbarian Europe where the Church held great power. The Church wielded a great deal of influence with the pagan Germanic tribes of the Danube River. The Romans had been warring with these German barbarians for a long time.

The Germans were known as the barbarians, or "bearded people," of Europe. They were sometimes also referred to as the "people of Ister." Ister was the Roman name for the Danube region of Germany. This name was originally derived from a celebration held there every spring by these pagan Germanic tribes, in order to guarantee the fertility of their crops. This celebration, called Ishter, (Babylonian Ishtar) was held every spring to honor Ishtar, the Babylonian goddess of fertility.

These blond-haired, blue-eyed people of the upper Danube, whose children possessed platinum-white hair, were thought to be direct descendants of the Caucasian "white-skinned" peoples of the Caucuses Mountains of southern Russia, called the "Russ." These "Russ," for whom Russia was originally named, migrated northward and westward across Europe to populate Germany, the Baltics, Scandinavia, Britannia, and extended their influence eastward across the steppes of Russia. Their tartan plaids, witches hats, and sturdy little ponies can still be found all the way from the northern plains of Mongolia, to the Shetland Islands in England.

These white-skinned people, who once lived under the rule of the ancient Babylonians, descended through ancient Media and the Caucuses Mountains of Russia into their new European homeland. Unfortunately they also carried their Babylonian gods and customs along with them. Their path through history can be traced by the peculiar habit of their males of wearing skirts. They are recorded as wearing these skirts when they were with the Medes in what is now Armenia (Ar-Media), and again many years later in the Danube region of Germany as the "Sigynnae," by the Greek historian Herodotus.

The Ishter holiday was celebrated by these Germanic tribes every spring, and was based upon astronomical positions of the sun and moon. The Ishter holiday was always celebrated on the first Sunday after the first full moon, after the spring equinox of the sun. This astronomical date, originally set by ancient Babylonian astrologers, is still the date on which Easter is celebrated today.

The original Easter (Ister) celebration was held all over northern Europe by pagans seeking to guarantee the fertility of their summer crops by paying tribute to Ishtar, the Babylonian goddess of fertility. The Ishter celebration was held in every village and town, and included the consumption of copious quantities of beer and wine by the male elders of each village, who would then seek out the village "virgins" for a little fun and merriment.

This "virgin" portion of the celebration was marked by a formal dance around the village May-Pole by unmarried females, who would dance in a provocative manner in order to stimulate the interest of their male suitors.

When the May-Pole dance was completed, the virgins would flee into the woods, pursued by their drunken male suitors. This reverse form of "Sadie Hawkins Day" ® often established new bonds for future marriages in the village. The many familiar Easter fertility symbols were also a necessary part of every Ishter celebration.

Rabbits and chickens, noted for their prolific propagation abilities, were ritually sacrificed to Ishtar, daughter of Sin, the Babylonian Moon-god, and then consumed by all the celebrants. Children were sent into the forest to hunt for decorated Easter eggs, and usually got to eat and enjoy their prizes. It was also the pagans' custom to bake cake offerings to the "Queen of Heaven" (Ishtar)(Jer.7:18). These little cakes were the forerunners of our modern "Hot Cross" buns, baked by modern Christians for their Easter celebrations.

Many modern libraries still contain paintings of these early Easter orgies. If you've ever wondered why European males are pictured wearing tights, these paintings leave no doubt that their purpose was to accentuate the lower part of the male anatomy, just as modern-day clothing is designed to accentuate the upper portion of the female anatomy. In those days, European males were apparently not very shy about their maleness.

When Catholic priests first ventured into the forests of Europe and witnessed these pagan orgies, they tried to discourage the sexual orientation of the celebration, but met with heavy resistance from village elders. They decided instead to negotiate with the pagans in order to lend a Christian air to the festivities.

Since Ishter was held every spring to celebrate the resurrection of green plants from their winter death, it was agreed that the celebration would also be used to commemorate the resurrection of Christ from death.

The Catholic priests found this compromise technique to be quite useful in extending their influence into the farthest reaches of the empire. Prevailing opinion was that the village elders would eventually die off, thereby allowing the priests to educate the younger generation into accepting the principles of Christianity.

This plan however, never quite materialized, and today the Church still retains all the original pagan fertility symbols, including rabbits, chicks, and eggs, in its Easter celebrations. Easter sunrise services are also still a vital part of the Easter morning mass, reflecting pagan worship of the sun.

Our Christian God is an invisible God, but the pagans needed a God that they could both see and touch. They therefore steadfastly refused to give up their idols and statues. One statue that was central to the Ishter celebration was the statue of the Babylonian goddess Ishtar, represented by the naked, pregnant female body. This statue celebrated the female anatomy in its most fertile state.

Catholic priests quickly took advantage of this statue by telling the pagans that it could be used to represent the pregnant body of the Virgin Mary, mother of Jesus. This change in focus from Jesus Christ to the Virgin Mary ultimately resulted in the early Church often being referred to as the "cult of Mary."

All in all, the Catholic priests were quite successful at compromising their way into pagan philosophy. They were also successful in their efforts to turn the pagan celebration of Ishter (L. Ister) into our modern-day Easter celebration. This compromise technique was also used on the pagan Romans as well. Through many such compromises, the Catholic Church was ultimately able to place itself into a very powerful position in Europe that guaranteed its existence for the next 1700 years.

The prophets tell us that God does not permit His people to compromise, and that it is every Christian's responsibility to continually examine his personal behavior, and to make sure it conforms to God's strict laws. Down through history, God's people strayed from Him many times and took up the worship of pagan gods and idols. In every case, they were severely punished for their misbehavior.

Many Christians celebrating Easter today probably do not realize that their Easter celebration is rooted in pagan Babylonian traditions. Christians however, by their own tenets, are expected to study the Holy Scriptures, and to be wary of any pagan influence in their religious celebrations.

CHAPTER TWENTY-TWO: OUR LADY OF FATIMA

"Know that this is the great sign given you by God that he is about to punish the world for its crimes, by means of war, famine, and persecutions of the church..."

HISTORY RECORDS MANY INTERESTING and prophetic events. Some of these events are recorded for us in the Bible, but some have also occurred in modern times as well. Often these modern prophetic events come under intense public scrutiny, and a few have even been exposed as frauds. Occasionally however, an event occurs that defies all reasonable explanation, and thus gains credibility with the masses. The following story recounts just such an event.

In the year 1916, a prophetic event occurred in Europe that was destined to affect the lives of the world's Catholics for the next century. In the tiny village of Fatima, Portugal, three young children, Lucia DeJesus, 9 years old, and her two younger cousins, Francisco and Jacinta Marto, ages 8 and 6 respectively, witnessed the appearance of an angel who identified himself as the Guardian Angel of Portugal. This angel appeared to the children a total of three times, and asked them to pray for more angelic appearances.

The children, captivated by the appearance of this angel, dutifully prayed as the angel had requested, and the following year another angel appeared to them as they tended their sheep in a field near their small village. The children described this second angel as a beautiful lady who informed them that the present war, World War I, would soon be over, but that a second and even more terrible war would eventually follow. The angel lamented to the children that the people of the world were not obeying God's laws, and therefore must suffer God's vengeance in the form of another world war.

The first time the children witnessed the appearance of "Our Lady" was on May 13th, 1917, followed by a second apparition on June 13th. Word of these miraculous visions spread quickly, and when the children visited the same field on July 13th, they were accompanied by nearly 5000 curious onlookers. Local officials however, were becoming concerned about the large numbers of people flowing into the small town of Fatima, and in August the children were placed under arrest to prevent a public panic.

An August vision occurred anyway on the 19th of the month, followed by further apparitions on September 13th and October 13th. By the time of the October vision, the crowds of onlookers had grown to more than 50,000 people. Many people reported witnessing miraculous astronomical occurrences, including the sun cavorting about in the sky, and various colors appearing in the heavens.

"Our Lady" revealed a total of three prophecies to the children concerning future events. The story of these miraculous visions eventually found its way to the world's Catholics, and the tiny town of Fatima, Portugal became an object of great interest for Catholics all around the globe. Ten-year old Lucia was the main source of information concerning the prophecies. She dutifully related all that she had been told by "Our Lady" to a local priest, and the three Secrets of Fatima were eventually delivered into the vaults of the Vatican Library, to be examined for authenticity. They were also submitted for official acceptance to Catholic Church hierarchy. The Secrets of Fatima could not be released to the general public without the express permission of the Pope himself.

The First Secret given to the children concerned a vision of human souls suffering in the fires of hell. Ten-year old Lucia describes what she saw in this first vision.

"Our Lady showed us a sea of fire, which seemed to be under the Earth. Immersed in this fire were many demons and souls in human form, like transparent burning embers, all blackened or burnished bronze, floating about in the conflagration, now raised into the air by flames that issued from within them together with great clouds of smoke; now falling back on every side like sparks in a huge fire without weight or equilibrium, amid shrieks and groans of pain and despair which horrified us and made us tremble with fear. The demons could be distinguished by their terrifying and repulsive likeness to frightful and unknown animals, all black and transparent. The vision lasted for but an instant. How can we ever thank our kind, heavenly mother who had already prepared us by promising in the first apparition to take us to Heaven, otherwise I think we would have died of fear and terror."

This frightening vision was enough to scare the living daylights out of the most ardent of sinners. The thought of spending an eternity in this kind of hell would make the trials of everyday life seem a blessing.

The Second Secret of Fatima was thought to contain a warning about the coming of a second worldwide war. Lucia was told that if men continued to offend God, a second, and even more terrible war would occur during the pontificate of Pius XI. She was told that a strange light would appear in the skies over Europe, and that God would then punish the world with war, hunger and pestilence, even punishing the Church and the Pope himself. During the First World War, in addition to the 9 million war deaths, a great pestilence known as the Spanish Flu spread across the world, killing over twice as many people as had died in the war. Unfortunately, young Francisco and Jacinta were destined to become victims of this great pestilence.

This vision of Catholics burning in hell, and another of God punishing the Church and even the Pope, did not sit well with the Catholic hierarchy, who much preferred a God that forgave everyone for all their sins. The idea that God might punish the Church, and even the Pope himself, did not do much to further the cause of the Church's acceptance of the three Fatima visions. The fame of Fatima however, continued to grow steadily, and ultimately the Church was obliged to place its seal of approval upon the Fatima visions in 1930.

The Second Secret of Fatima was rumored to contain a warning about another worldwide war that might occur sometime in the future. The Second Secret was officially released from the vaults of the Vatican in 1941. Lucia describes this second vision given to her by Our Lady.

"You have seen hell where the poor sinners go. To save them, God wishes to establish in the world a devotion to my Immaculate Heart. If what I say is done, many souls will be saved and there will be peace. The war (World War One) is going to end soon, but if people do not cease offending God, a worse war will break out during the pontificate of Pius XI. When you see a night illuminated by an unknown light, know that this is the great sign given by God that he is about to punish the world for its crimes by means of war, famine and persecutions of the Church and of the Holy Father. To prevent this, I come to ask for the consecration of Russia to my Immaculate Heart, and the Commission of repentance on the First Saturdays. If my requests are heeded, Russia will be converted and there will be peace, if not, she will spread her errors throughout the world causing wars and persecutions of the Church. The good will be martyred, the Holy Father will have much to suffer, and various nations will be annihilated. In the end my Immaculate Heart will triumph. The Holy Father will consecrate Russia to me, and she shall be converted; a period of peace will then be granted to the world."

On January 25th, 1938, an unusually large Aurora Borealis was witnessed over the continent of Europe. Thousands of people telephoned authorities wondering if this was the end of the world. A short time later, Adolph Hitler invaded Austria. Then, on January 25th, 1939, exactly one year after the lights illuminated Europe's skies, Hitler issued his Jewish Policy edict, and the Second World War was officially on.

Lucia made a decision to go into the service of the church as a nun, and it was decided that no more secrets would be released until a much later date. In the 1950's, the Prophecies of Fatima were a primary focus of attention for the world's Catholics, who desperately wanted to prevent a future war with atheist Russia. Shrines to "Our Lady of Fatima" were constructed in Churches all over the world as places where Catholics could go to light candles and pray for the conversion of Russia. Catholics everywhere dutifully prayed for this conversion. Every Catholic was thoroughly familiar with the prophecies of Fatima, and the unreleased Third Secret was still the subject of much anxious discussion in many religious circles.

The Third Secret of Fatima was rumored to contain a reference to the assassination of a pope. This Third Secret was scheduled to be opened and read by the Pope in 1960; but when Pope John XXIII opened and read the prophecy, he was so horrified by it that he refused to release it. He placed the Third Secret of Fatima back into the Vatican's vaults, and the next 40 years witnessed interest in Fatima fading from the minds of most Catholics.

Then, in May of the year 2000, Pope John Paul II met with Sister Lucia at the Vatican, and decided that it was finally time for the Third Prophecy of Fatima to finally be released. An official Church statement released along with the prophecy proclaimed it to be a "figurative representation" of the attempted assassination of Pope John Paul II that had taken place on May 5th, 1981. It should be noted however, that the appearance of the great light in the skies over Europe was not a figurative event; and the second prophecy concerning Russia's conversion from Communism, was also not a figurative event.

The Third Prophecy of Fatima describes a Vatican in ruins, littered with human corpses, and a Pope held prisoner by the soldiers of an invading army. These soldiers march the Pope and his bishops to the top of a hill to pray in front of an old rugged cross. The soldiers then execute the Pope along with his bishops and a host of others. This haunting vision is very similar to one experienced by an earlier pope in the year 1910.

Pope Leo X is said to have awakened suddenly from a dream and announced that he had just seen a horrible vision of a future pope exiting the Vatican over the dead bodies of his bishops.

This vision also matches that of Saint Malachi O'Morgan, a 12th century Roman Catholic priest from Ireland, who described the coming of Peter the Roman, last Pope of the Roman Catholic Church. Peter witnesses the destruction of the seven-hilled city (Rome), followed by the arrival of the Dreadful Judge, who arrives to judge the nations of the world. Malachi places the papacy of Peter the Roman after that of the successor to Pope John Paul II.

Malachi's vision, if true, places us only two pontificates from the fulfillment of the Third Fatima prophecy. Many Church leaders in America also insist that this exact prophecy is also described within the pages of the book of Revelation. It is thought to be the last event before the Lord's return.

This is Lucia's description of the Third Secret as revealed to her by Our Lady so long ago in that field near her village.

"At the left of Our Lady and a little above, we saw an angel with a flaming sword in his left hand, flashing, it gave out flames that looked as though they could set the world on fire, but they died out in contact with the splendor Our Lady radiated toward him from her right hand. Pointing to the earth with his right hand, the angel cried out in a loud voice, "Penance! Penance! Penance!," and we saw a great light that is God, something similar to how people look in a mirror when they pass in front of it. A bishop dressed in white, we had the impression it was the Holy Father, and other bishops, priests, and religious men and women were going up a steep hill, at the top of which was a big cross made of rough tree trunks, like a cork tree with its bark. Before reaching there, the Holy father passed through a big city, half in ruins and, trembling, in halting steps, suffering from deep pain and sorrow, he prayed for the souls of the corpses he passed along the way. Having reached the top of the hill, he knelt down at the foot of the big cross, and was killed by a group of soldiers who fired upon him. In the same way died all the other bishops, priests, and religious men and women, and also various lay people of different rank and position. Beneath the arms of the cross stood two angels, each with a crystal aspertorium in his hand, with which they each gathered up the blood of the martyrs, and with it sprinkled the souls that were making their way to God."

Could this chilling vision actually come true? The First World War began with the assassination of the heir to the Holy Roman Empire on July 28th, 1914, when Archduke Franz Ferdinand was assassinated while riding in an open car through the streets of Sarajevo. It has also been predicted that the last world war will be initiated by the assassination of the last Pope of the

Holy Roman Church. It is very strange indeed that a prophecy such as this should come from the Vatican's own vaults.

CHAPTER TWENTY-THREE: ANGOLMOIS

(Q. 6-100) THE LAW OF PROPHECY AGAINST INEPT INTERPRETERS: "Let whoever reads my verses think carefully. Let the impious and the ignorant not attempt them. Seers, idiot astrologers, the simple-minded, and evildoers, stay way! Whoever does otherwise, he rightly, is accursed!

Michel deNostradame

THE MOST FAMOUS PROPHECY in Nostradamus' entire collection is a prophecy that came to be known as the "King of Terror" prophecy. It was unique from all the other prophecies in Nostradamus' famous collection, because it contained an exact date for the prophecy's' occurrence.

Prophecy #10-72 was actually an experiment set up by Nostradamus to see if anyone in this modern day and age would be smart enough to figure out one of his prophecies before it occurred. It was a challenge to the greatest minds of the 20th century.

This prophecy, found in Chapter 10, Quatrain 72, of the *Centuries*, was the only one of his more than 900 prophecies that actually contained both the month and year that the prophecy was to take place. The date was July 1999. If this prophecy came true, it would be the final proof needed to dispel the doubts of unbelievers everywhere. All this created quite a stir in the world of prophecy just prior to the 2000 Millenium.

The arrival of a millenium always generated a great amount of anticipation and fear in the common people. The arrival of the previous millenium in the year 1000 AD, created great panic in Europe, as Christians in all over Europe and Asia nervously anticipated the arrival of the great Day of Judgement, and the end of the world.

As was also the case in the year 2000, there was no shortage of those aiming to profit from the widespread panic. The Holy Roman Church encouraged its members to do everything they could to improve their standing with the Lord as the dreaded Judgement Day approached. European Christians, and Byzantines from the Eastern Church, flocked by the thousands into their great cathedrals carrying armloads of gifts of money, jewelry, deeds for land, and wagonloads of valuables to fill church coffers.

On December 31st, 999 AD, peasants and landlords alike from all over the Christian world, crammed into their churches and cathedrals to await the arrival of the dreaded Day of Judgement. When the midnight hour passed with no Judgement occurring, the church suddenly announced that the offerings had apparently been sufficient to forestall the great Day of Judgement, and the Church was able to erect many new churches with all the money and gifts it had gained during the great fiasco.

As with previous milleniums, the 2000 Millenium also generated similar fear and panic. The media joined in the frenzy, and millions of dollars were spent as the world prepared for something called "Y2K," which was supposed to shut down the entire computerized world at the stroke of midnight on December 31st.

As Nostradamus' dreaded July 1999 date grew nearer, interest in his prophecies also increased, and more Nostradamus books were sold at this time than at any other time in history. The original prophecy #10-72 was a four-line poem known as a quatrain. A rough English translation of the quatrain read as follows:

CHAPTER 10, QUATRAIN 72 (English translation)

In the year 1999, seventh month,

From the sky, arrives a powerful leader

To resurrect the great king of the Angolmois.

Before this, war reigns for a good cause.

The key to the correct interpretation of this prophecy was the word Angolmois. Nostradamus was known to anagram certain words or names in order to disguise their meaning, and Nostradamus buffs decided that the word Angolmois was actually an anagram of the word Mongolois. They therefore concluded that a great Mongolian antichrist was going to come down out of Asia to invade the Middle East, setting off the final battle of Armageddon. This particular interpretation of the prophecy fit in perfectly with all the other ominous predictions of the time.

When the month of July finally arrived, Nostradamus followers all over the world anxiously awaited this Mongolian antichrist's invasion of the Middle East. But the first week of July passed with no unusual events. Then the second and third weeks of July also passed without incident. When the entire month of July passed without the appearance of this Mongolian antichrist, Nostradamus followers were shocked and disappointed. They just couldn't believe that nothing at all had happened. They quickly came up with all sorts of excuses to explain their embarrassment, not realizing that the prophecy actually had come true.

The word Angolmois you see, was not really an anagram after all. The word Angolmois in Old French simply means "Angol-people," but who in the world are the Angol people? The word Angol, spelled A-N-G-O-L, does not exist in any known language. There are no Latin or Greek roots for this word. The only word in any language that contains the exact spelling A-N-G-O-L is the word Angola. Angola is a Portuguese trading colony on the western coast of Africa.

In the late 1400's the Portuguese established a trading relationship with a nomadic African tribe located in present-day Angola. They traded for such things as ivory, animal skins, gold and diamonds. The leader of this nomadic tribe was called the "n'gola," or king. Thus the name "Angola" was given to the colony. So in Old French, the word Angolmois actually means "Portuguese African Kingdom." These Angolan nomads later moved north and established a second trading colony known as Morocco, located much closer to Portugal and therefore more convenient for trading, thereby creating two Portuguese "angolas," one known as Portuguese Angola, and the other, Portuguese Morocco.

Nostradamus' prophecy referred to a leader of awesome power who would arrive out of the sky into one of these two Portuguese colonies and witness the resurrection of a king upon the throne of his native people. But which colony, Angola or Morocco? And who was this powerful world leader?

Well, the world's most powerful leader in the year 1999 was America's president, Bill Clinton. This meant that prophecy #10-72 was actually predicting that in July of 1999, President Clinton would arrive out of the sky into either Angola or Morocco to witness the coronation of a king of one of these two

African nations. The last line of the prophecy mentioned that the President would also acknowledge a war that was fought for a "good" cause.

Well folks, believe it or not, on July 24th, 1999, President Clinton actually did arrive out of the sky on Air Force One into the former Portuguese trading colony of Morocco to witness the resurrection of Morocco's Prince Hussan on the throne of his deceased father, King Hussan. After witnessing Hussan's coronation on the throne of his people, President Clinton stepped back into Air Force One and flew off to Kosovo to recognize the bravery of American soldiers who'd just fought in a war for the "good" cause of preventing a human genocide.

President Clinton's trip to Morocco was totally unplanned. It was an unanticipated side trip prompted by the unexpected death of King Hussan, to whom Clinton owed a debt of gratitude because the king had been instrumental in negotiating an earlier Arab-Israeli peace agreement.

This truly incredible feat of modern prophecy went totally unnoticed and uncredited, even though it came true exactly as Nostradamus predicted it would.

Nostradamus' followers, by incorrectly interpreting the prophecy, had discredited the great prophet. The scientific world may scoff at the world of prophecy, but the following Quatrain can be found in numerous books in many libraries and bookstores all over the world. And all of these books were published many years in advance of the year 1999. For your convenience, I've included a glossary of the Old French and Latin terms for your reference.

Chapter 10, Quatrain 72 (Old French)

L'an mil neuf cens nonante neuf, sept mois,
In the year nineteen hundred ninety nine, seventh month,

Angolmois

Du ceil viendra un grand Roi de-ffraieur
From the sky will come a grand leader of (great) power

Resusiter le grand Roy d'Angolmois.
To resurrect the great King of the Angol people

Avant que, Mars regner par bon-heur.
Before this, war reigns for a good-cause (good-time)

OLD FRENCH AND LATIN DEFINITIONS:
Angolmois – (O.F.) Angol people
Avant – (F.) before
Bon-heur – (O.F.) good-hour, good cause
Ceil – (O.F) sky
Cens – (O.F) hundred
De-ffraieur – (O.F. frayeur) frightening power
Grand – (F.) grand, great
L'an- (O.F. an - year), in the year
Mars – (O.F. Mars, god of war) war
Mil – (O.F. mille) one thousand
Mois – (F.) month
Neuf – (F.) nine
Nonante – (L. nonagin) ninety
Par – (F.) for
Que – (F.) this, that
Regner – (F.) reign
Ressusiter – (O.F.) resuscitate, resurrect
Roi, Roy – (O.F) leader, King
Sept – (F.) seventh
Viendra – (O. F. venir, viendra) will come

This quatrain represents a truly an incredible feat of prophecy by the modern Hebrew prophet Nostradamus. It is sad to note however, that Nostradamus' great experiment also proved that men of the 20th century were not intelligent enough to figure out one of his prophecies before it occurred, certainly not much of a tribute to the IQ of so-called "modern" man!

CHAPTER TWENTY-FOUR:
THE STONE OF DESTINY

And Jacob rose up early in the morning, and took the stone that he had used for his pillow, and set it up for a pillar, and poured oil upon the top of it.

(Gen. 28:18)

FOR CENTURIES, MEN HAVE BEEN CAPTIVATED by tales of mysterious holy relics such as the Ark of the Covenant and the Holy Grail. These objects spawned numerous books and movies about their magical powers and prophetic mystique. The existence of these holy objects however, can only be confirmed by ancient legend. None of these objects has ever been located. There is however, one holy item that has survived into modern times. It is the famous Stone of Destiny. This holy object predates all others, and carries with it a history and a destiny that is truly the stuff that great legends are made of.

The significance of the stone is deeply rooted in Bible history, for the stone is thought to carry God's blessing upon those who possess it, and it was destined to follow God's people on their long journey through time. The story behind the Stone of Destiny is nearly 4000 years old, and I therefore thought you might enjoy reading about this holy object that really does exist, and still has a role to play in modern history.

In order to discuss the stone however, it will first be necessary for us to discuss the blessing that it carries. We all know that in the beginning God created Adam, and that the bloodline of Adam still exists upon Earth today. This Genesis story comes down to us through the ancient Hebrew Scriptures. These Scriptures also tell us that in the time of Adam, the sons of God gazed upon the daughters of men and saw that they were fair; and so the sons of God went in unto the daughters of men and bare them children. Thus began the line of the great patriarchs, a superior race of men created through this strange union. These biblical patriarchs had the unique ability to live for ten times as long as normal human beings, but unfortunately they also carried with them the carnal nature of man.

The world of the patriarchs eventually developed into a world of great evil, and so God decided to destroy what He had created. There was however one patriarch named Noah, who was blessed with goodness, and so God decided to save Noah and his family from the destruction of the great Flood. We all know the familiar story of Noah's Ark and the Flood, and how Noah and his family survived this great catastrophe. Noah was the first in a blessed line of men to carry God's blessing into the modern world. God's blessing was passed down from father to son along a genealogical line that is carefully recorded for us within the pages of the Bible.

After approximately thirteen more generations, God appeared to the prophet Abram and promised Abram that he would eventually father a great nation. Abram's name was then changed to Abraham (father of a multitude), for he would father a blessed line of the Hebrew people. It is this same blessing that the stone still carries with it today.

The Stone of Destiny, also called "Jacob's Pillow," was just another stone lying on the desert floor until one night in 1900

BC, when Jacob used it as a pillow while on a trip from Harab to Bethel (Gen. 28:18). While sleeping on the stone, Jacob dreamed of a great ladder leading up to heaven with angels climbing up and down it. In the dream, the Lord speaks to Jacob, telling him that his children will be a blessing unto the world. When Jacob arose in the morning, he took the stone along with him, and thus the stone began its long trek through time.

Jacob's name was eventually changed to Israel (those of God), and he fathered twelve sons who would be the progenitors of the twelve tribes of Israel. Jacob could only bestow God's blessing upon one of his sons, and so he sat his twelve sons down one day, and told them of the ultimate destiny of their descendants in the last days (Gen. 49). Jacob decided to bestow God's blessing upon his son Joseph, and so the blessing, and the "Shepherd's Stone," would follow the descendants of Joseph until the last days (Gen. 49:24).

The location of the stone remained a mystery for many centuries. It was eventually traced to Great Britain, and now resides in Edinburgh Castle in Scotland. But because the Irish, the Scots, and the English so love to tell tall tales, there are quite a number of stories about how the stone first arrived in Great Britain. The most probable story however, involves the biblical prophet Jeremiah.

It is said that around the year 580 BC, when the Babylonians sacked Jerusalem, Jeremiah removed the stone from the Great Temple in Jerusalem and brought it to Ireland by sea, along with Tea Tephi, daughter of King Zedikiah of Judah. The stone was then used to wed Tea Tephi to the Irish king, Eochaide the Heremon, thus preserving the bloodline of David upon a world throne.

From that day forward, all Irish Kings were required to be enthroned in the presence of the stone. The stone was kept at Tarah, in Meath, and followed the line of the Dalriadic kings down through Irish history. Even Saint Patrick is said to have blessed this stone that followed the descendants of Erc, first king of the Dal Riata tribe of Antrim.

Around the year 850 AD, Fergus is rumored to have borrowed the "Lia Fail" (Stone of Fate) from his brother, King Muisceortagh of Ireland, in order to use it at his coronation in Alba (Scotland). Fergus however, never returned the stone to his brother. The stone remained in Scotland, and eventually found a home in the city of Scone (pronounced "scoon"). This "Stone of Scone" was then used to enthrone all subsequent kings of Scotland. The last Scottish king to be enthroned in the presence of the stone was John Baloil in the year 1292.

When Edward I of England invaded Scotland in 1296, Baloil surrender the stone to Edward, and abdicated the Scottish throne. Edward then transferred the stone from Scone to Westminster in England.

In 1301 the English built a "Coronation Chair" to house the stone, and this chair was thenceforth used to enthrone all subsequent British monarchs.

Then, in the year 1328, the Scots formally won their independence from England with the signing of the Treaty of Northampton by Edward III, who promised to return the stone to Scotland. The stone however, was never returned. And so on Christmas day in 1950, a group of Scottish students from Glasgow decided to reclaim the stone from Westminster Abbey and return it to Scotland. They broke into Westminster Abby and removed the stone from beneath its Coronation Chair. They then placed the stone in the trunk of their car and spirited it off to an unknown location.

Scotland Yard finally located the stone many months later in Arbroath Abbey in Scotland. It was formally reinstated in Westminster Abbey in February of 1952, and in 1953 was used at the coronation ceremony of Queen Elizabeth II. The stone still remained a source of controversy however, and in 1996 the

Queen finally agreed to allow the stone's return to Scotland, with the stipulation that it be made available for any future coronations.

According to ancient prophecy, the Stone of Destiny will follow an unbroken line of monarchs from the throne of David until Shiloh (the Lord's return). We know from the record of history that this sacred stone was indeed used to enthrone a long line of Irish, Scottish, and English monarchs from 580 BC to 2000 AD, and may assume that the prophet Jeremiah delivered it to Great Britain for this express purpose. This is a truly amazing case of ancient prophecy still operating in modern times.

There are many other curious links between the ancient Hebrews and the people of the British Isles, particularly the Scottish people. It seems that the ancient Hebrews were also known to wear skirts, or kilts, whose woven plaids displayed the colors of their various tribes. Also, the bagpipes now used by Irish and Scottish pipers are almost identical to the ancient Hebrew goatskin bagpipe that David played while tending his sheep. And ancient Hebrew masonic guilds were employed by other civilizations whenever stone buildings or monuments needed to be erected. There are some who insist that this masonic link extends to the original formation of the United States of America in 1776. The famed Washington Monument that sits at the very center of our Capital Mall in Washington DC, is in fact a masonic obelisk, and was erected by the members of the guild of the ancient Scottish rite.

There are some who still doubt that the British Stone of Destiny is in fact Jacob's Pillow, but the British Crown continues to observe this ancient rite. This is just one more case of ancient prophecy that is still operating in modern times.

CHAPTER TWENTY-FIVE: ASTEROID

"That day is impending when people will admit the pure truth within the book of nature, as well as in the Holy Bible, and rejoice at the harmony between these two revelations."

Johannes Kepler

SOME PEOPLE SAY THEY GO THROUGH LIFE feeling like a duck in a shooting gallery. Well, this statement might be truer than many of us realize. In some earlier chapters of this book we established that America's early colonists believed that God controlled the weather and all other natural events, and that God often used these natural disasters to deliver His vengeance upon mankind whenever men strayed from obeying His laws.

We later learned from the Fatima prophecies that God also uses war and pestilence to deliver similar punishments upon man. The largest and most devastating natural disaster of all however, was not mentioned in any of our previous stories. It's an event that has been the subject of some best-selling books and videos in recent years. This most frightening of all natural disasters is the asteroid strike.

The idea that an immense asteroid might suddenly and unexpectedly strike Earth, has been a frightening reality for scientists and astronomers for many years. Outer space is filled with many small objects that spin around in our solar system in widely eccentric orbits, and some of these objects occasionally pass very close to our planet.

Most of the movies made about these giant space rocks involve plots concerning someone saving the Earth from certain disaster by sending out nuclear missiles to blow up the asteroid or knock it off course. Just how real is the possibility that an asteroid might strike the Earth? Well, let's take a closer look at that question to see if we can find the answer.

There are two different kinds of roving objects that normally travel through space and threaten the planets of our solar system; they are comets and asteroids. Comets are huge balls of loosely packed space debris that move rapidly through space and give off light due to the reflection of the sun off their vapor trails. Comets are therefore quite easy for us to spot if they approach Earth. Since comets are so easily seen, we would at least have some time to react if a comet were to pose any real danger to our planet.

And just how dangerous are comets? Well, in the summer of 1908, a small comet exploded in the air over the forests of Tunguska, Siberia, causing massive devastation, and felling trees in a circle for 20 miles in every direction from the epicenter of the explosion. It is estimated that this small comet had a diameter of less than 150 feet, and was traveling at a speed of approximately 40 miles per second. It was however, able to generate an explosive force equivalent to about 10 times that of the Hiroshima bomb. That's right, it was like setting off 10 Hiroshima nukes in one place at the same time.

Luckily, this tiny comet struck in one of the most desolate places on our planet. If it had struck near a major city, it would have exploded with a force capable of destroying the city and its entire population.

Trail of Prophecy

In 1994, Earth's astronomers witnessed a large comet strike the surface of the planet Jupiter. This comet, called Shoemaker-Levy 9, was captured by Jupiter's gravitational field and orbited the planet briefly, before breaking up into 21 pieces and crashing into the planet's surface. The largest piece of this comet was estimated to be almost 2 miles in diameter. Scientists said that if this same comet had crashed into Earth, it would have destroyed all life on our planet. The planet Jupiter however, is a thousand times the size of Earth, and therefore did not suffer unduly from the collision.

Another space object that poses a much greater danger to us, is the asteroid. The reason that asteroids are so dangerous is that they do not give off any light, and are therefore very difficult to see in the blackness of space. Asteroids may therefore strike us without warning.

The myriad of craters visible on the surface of the Moon, were created over many millennia by thousands of asteroid strikes. The Moon has no atmosphere with which to generate the rain and wind necessary to erode these craters. The Earth has been struck just as often as the Moon by asteroids from outer space, but most of the craters created by these many strikes have been completely eroded, due to the effects of wind and water.

Asteroids, or meteors as they are sometimes called, have struck Earth many times. Actually, hundreds of tiny meteors strike the Earth every day. They are the familiar shooting stars that we see each evening in the sky. Very few meteors are large enough to punch through Earth's atmosphere and reach the ground intact, but some do.

In modern history there have been a few celebrated incidents involving meteor strikes. The one time a meteor is recorded to have struck a person was in 1954 in Alabama. A meteor weighing approximately 8 pounds crashed through the roof of a woman's home, ripped through her ceiling, and bounced off her console radio, eventually striking her on the hip while she was seated on her couch, leaving her with a large bruise for the experience.

Another famous incident involved an amazing coincidence where two separate meteor strikes occurred in two homes located only about a mile apart in the town of Weathersfield, Connecticut. One home was struck in 1971, and the other was struck over a decade later in 1982.

Then, on the night of October 9th, 1992, a giant fireball was witnessed in the skies over the town of Peekskill, New York. When a local resident heard a loud crash outside her house, she went outside to investigate the noise, and found that a meteor had pierced the trunk of her daughter's Chevy Malibu that was parked in the driveway, and left a gaping hole in the pavement underneath the car.

If you trace this woman's family name through the genealogy websites, her name oddly enough can be traced back to its original roots in the 1600's in the small town of......you guessed it.....Weathersfield, Connecticut!

In the many movies and videos created about large asteroid strikes, Earth's scientists generally discover the asteroid as it is approaching Earth, and sound a general alarm. The government then initiates a hurried plan to destroy the asteroid before it crashes into our planet.

In real life however, this situation is not quite as secure as described in these science fiction productions. Actually the chances of us ever sighting an asteroid in time to plan a coordinated response, are extremely small.

In March of 1989, a 1500-foot asteroid just missed Earth, and was not spotted until it was 650,000 miles past our planet, going away! If this asteroid had struck us, it would have delivered a punch equivalent to 3000 nuclear warheads!

On June 14th, 2002, another asteroid, about 300 feet in diameter, passed by about one-fifth the distance to the Moon, and also was not spotted until it was going away.

Since asteroids travel at speeds of about 40 miles per second, they can cover a distance of 500,000 miles in as little as 3 or 4 hours. Even if we were lucky enough to spot one of these asteroids at this great distance, there would be little time for us to react.

In 1932, 1936, and 1937, three huge asteroids named Apollo, Adonis and Hermes, all passed within about 500,000 miles of our planet. These asteroids were true giants, each one packing an explosive force roughly equivalent to 50,000 nuclear warheads. If one of these asteroids had struck our planet, it could have marked the end of civilization as we know it. Oddly enough, it seems that scientists in those days were much more watchful of our skies than modern-day scientists. The Hermes asteroid was spotted by astronomers in Heidelburg, Germany many days before its arrival.

Over time, these huge asteroids have been observed to strike Earth with a predictable frequency. Small asteroids tend to strike the Earth quite often, while larger asteroids usually strike only about once every 50 million years or so. The chart below will give you some idea of the explosive force and strike frequency of these huge space monsters.

DIAM.	MEGATONS	STRIKE FREQ.	CRATER
200 ft.	10	1,000 yrs.	10,000 ft.
1500 ft.	3,000	10,000 yrs.	5 miles
½ mile	30,000	100,000 yrs.	15 miles
1 ½ mi.	500,000	1,000,000 yrs.	30 miles
8 miles	60,000,000	100,000,000 yrs.	80 miles

Asteroid

Approximately 65 million years ago our planet was struck by an asteroid almost 8 miles in diameter. This collision is thought to have been responsible for the extinction of Earth's dinosaurs. This asteroid strike created so much dust in Earth's atmosphere, that the planet suffered over three years of winter. The large, cold-blooded dinosaurs could not easily adapt to this sudden shift in temperature, and quickly succumbed to the freezing temperatures. Earth's warm-blooded creatures however, were better equipped to handle this abrupt temperature change, and therefore survived to re-populate the planet.

Since Earth is estimated to be about 3 billion years old, we can assume that it has been struck 40 or 50 times by asteroids this size during its long existence.

Asteroid strikes capable of causing major changes in Earth's climate are rare, but as recently as 500 years ago, there was a major climate change produced by an asteroid strike in the Pacific ocean just east of Australia. This asteroid was approximately a quarter of a mile in diameter, and its effects were felt all around the world.

Every moment that ticks by places us one step closer to another asteroid strike. It is just a matter of time until one of these giant space rocks arrives to wreak havoc upon our world as we know it.

In God's great arsenal, there is no natural force that compares to the asteroid. If a truly large asteroid (1 mile across) were to land in one of our oceans, it would create a tidal wave large enough to wipe out many coastal cities, and if it struck land, it could throw our planet into an Ice Age. We would be unable to grow any food, or heat our homes in this type of situation, and our cities would almost certainly perish. Civilization as we know it, would probably cease to exist.

Just how real is the danger of this sort of an asteroid strike? Well, in December of 1994, an asteroid of incredible proportions passed within a mere 60,000 miles of our planet (that's one-fifth the distance to the Moon). This is about as close to Armageddon as you can get. If this asteroid had crashed into Earth, you probably would not be sitting here reading this book today.

There is no force in God's universe that can match the destructive force of the mighty asteroid, and it is only through the grace of God that we have thus far avoided this threat to our very existence.

CHAPTER TWENTY-SIX: THE END OF TIME

And the Lord said, My spirit shall not always strive with man, for that he also is flesh: yet his days shall be an hundred and twenty years.

(Gen. 6:3)

WELL, WHERE DO WE GO FROM HERE? You've no doubt heard the saying that there are two things in life that can always be counted on, death and taxes.

Most of us take these two things for granted, and never question their inevitability. In fact, we spend most of our lives saving for retirement, and dutifully making all those monthly life insurance payments so our loved ones will be well taken care of when we're gone. But what if we didn't have to die? What if we could live forever? That's impossible, right? Well, don't be so sure about that until you first check with the Bible to see what it has to say on the subject. You might be surprised by what you find.

In the book of Genesis we read that sometime before the great Flood, God decided to fix the years of man at one hundred and twenty. This reference is found in Genesis Chapter 6, Verse 3. At the same time God fixed the age of man however, the biblical patriarchs were living for much longer than one hundred and twenty years. Noah in fact, lived to be 950 years old, and Adam lived to be 930!

The biblical patriarchs could live for ten times as long as normal human beings, but how was that possible? Death comes as a result of aging, doesn't it? We certainly can't stop the aging process, can we?

Well, it turns out that aging is caused by a tiny genetic time-clock that ticks away inside of every living human cell. This genetic timeclock is fixed within the genetic code of each and every one of our body cells, but remember, the patriarchs were not so very different from humans.

If you read a little deeper into the book of Genesis, you'll find that it tells the story of the beginning of the world, and also the story of the creation of man. It is also interesting to note that in the book of Genesis, written over 3400 years ago, the Bible tells us that man can only expect to live for a maximum of one hundred and twenty years. That's somewhat surprising, since at the time of Genesis the life expectancy of the average human being was less than 30 years. Was it mere coincidence that allowed the writer of the book of Genesis to get this fact correct?

Of course there are one or two people who claim to have lived for longer than one hundred and twenty years, but a thorough examination of their birth records leaves much to be desired. The overwhelming pattern of human longevity indicates a maximum life expectancy of exactly one hundred and twenty years.

The real question for mankind is, can we alter our genetic timeclocks so that we can live for longer than one hundred and twenty years? The book of Genesis does seem to indicate that this is possible.

When Adam and Eve were first created in the Garden of Eden, they were warned by God not to eat of the two trees in the center of the garden. One of these trees was known as the Tree of Knowledge, and the other, the Tree of Life.

Eve was duped by the Serpent into partaking of the fruit of the Tree of Knowledge, thus condemning mankind to 6000 years of attempting to create the perfect human society with that stolen knowledge. After Eve committed this act of original sin, God banished both Adam and Eve from the Garden so they would not also partake of the Tree of Life, and live forever as gods do (Gen. 3:22).

So the Bible does seem to indicate that it is possible for men to live forever if they eat from the Tree of Life. Now there's a frightening thought! Could it actually be possible for us to alter the genetic code within our cells and live forever as gods do? This intriguing thought brings up the subject of the curious genetic condition known as Progeria.

Progeria is an illness in which the genetic timeclock of the human cell has been altered. The word Progeria comes from the Greek *pro*, meaning "advanced," and *geria*, meaning "aging." Progeria is a genetic condition that causes premature aging in its victims.

Progeria sufferers can age at ten times the rate of normal human beings. They can actually die of old age when they are only ten years old!

In this rare genetic disease, the timeclock inside the human cell has been genetically altered, and ticks at almost ten times its normal rate. A Progeria victim who is only nine years old, appears in every way to be a ninety-year old person.

Only a few people on Earth suffer from this rare genetic disorder, and the condition occurs without respect to the race or gender of its victims. This disease however, provides us with living proof that it is possible to alter the genetic timeclock of the human cell.

Could it be possible that there are also some people on Earth who've had their genetic timeclocks altered in the opposite direction? We haven't heard from anyone with such a disease yet, but if you could live to be 1200 years old, you might not be telling anyone about it. After all, what would society do with all those unemployed life insurance salesmen?

Since the members of our scientific community seem to be so determined to fool around with our genetic codes, might it be possible that they will someday locate this particular code, and alter it?

This brings up another interesting thought. Perhaps the end of time, is merely the end of aging. Now wouldn't that be a kick in the pants!

BIBLIOGRAPHY

DANIEL AND THE REVELATION, Uriah Smith, 1897, Pacific Press Publishing Assoc., Mtn. View, CA

THE GREAT PYRAMID: WHY WAS IT BUILT & WHO BUILT IT, John Taylor, 1864, Longmans, Greene, London

HANDBOOK OF NORTH AMERICAN INDIANS, VOL. 15 NORTHEAST, Wm. Sturtevant, 1978, USA

THE LIFE OF PASTEUR, Rene Vallery-Radot, 1937, Sun Dial Press, NY

BENJAMIN FRANKLIN, Carl Van Doren, 1938, The Viking Press, NY

THE LIFE AND DEATH OF ADOLF HITLER, Robert Payne, 1973, Praeger, New York, Washington

THE BASIC WRITINGS OF THOMAS JEFFERSON, Philip S. Foner, 1944, Willey Book Company, NY

THE MAN WHO SAW TOMMORROW – THE PROPHECIES OF NOSTRADAMUS, Erika Cheetham, 1974, Berkley, NY

NAPOLEON, Emil Ludwig, 1926, Garden City Publishing, Garden City, NY

OUR INHERITANCE IN THE GREAT PYRAMID, Charles Piazzi Smith, 1864, Royal Astronomer, Scotland

THE PILGRIMS FIRST YEAR IN NEW ENGLAND, Nahum Gale, 1857, Boston

EVIDENCE FROM SCRIPTURE AND THE HISTORY OF THE SECOND COMING OF CHRIST. William Miller, 1842, Himes, Boston

www.ingramcontent.com/pod-product-compliance
Lightning Source LLC
Chambersburg PA
CBHW031832090426
42741CB00005B/211